Stormy Weather

A forecast of increased storms,
worsening weather and detailed
explanations of the causes

MARK MASLIN

APPLE

A QUARTO BOOK

First published in the United Kingdom
by the Apple Press
Sheridan House, 112–116a Western Road
Hove, East Sussex BN3 1DD

ISBN 1–84092–378–4

Conceived, designed and produced by
Quarto Publishing plc
The Old Brewery
6 Blundell Street
London N7 9BH

Project Editor Vicky Weber
Senior Art Editor Sally Bond
Assistant Art Director Penny Cobb
Designer James Lawrence
Copy Editors Gillian Kemp, Sarah Hoggett
Illustrators Andrew Green, Julian Baker, Tony Walter-Bellue
Picture Research Image Select International, Virginia Ettwein
Indexer Diana Le Core

Art director Moira Clinch
Publisher Piers Spence

Manufactured by Universal Graphics, Singapore
Printed by Midas Printing International Limited, China

QUAR. STOR

10 9 8 7 6 5 4 3 2 1

CONTENTS

FOREWORD

Notwithstanding an increasingly vicious rearguard action by the skeptical and the downright hostile, there is now no doubt that our planet is warming rapidly and that we are the cause.

Over the last few centuries carbon dioxide levels in the atmosphere—fed by rampant industrialization—have risen by about a third, cloaking the Earth in an insulating blanket and causing temperatures to rise by around 1°F (0.6°C) since 1900. And this is just the start. As the chimes herald the beginning of the 22nd century, the world could be roasting as a result of an increase in temperature ten times greater than this.

Nor will heat be the only problem. As the climate system struggles to adapt to one of the most rapid episodes of warming our planet has ever experienced, so it is likely that it will become increasingly unbalanced and erratic. The consequences are dire—flood, drought, storm, and wildfire. In the opening year of the new millennium, an incredible one person in every thirty was affected by natural hazards, many of which were weather related. As planetary warming accelerates, such statistics can only get worse.

It is highly likely that, as the oceans heat up, they will spawn more and bigger tropical cyclones, threatening coastal regions from wealthy Miami and the US Gulf Coast to the crowded and poverty-stricken coastal plains of India and Bangladesh.

As precipitation becomes more erratic, floods will deluge some countries while others will suffer interminable drought. By the middle of the century, perhaps five billion people will live in countries short of potable water, leading to conflict and economic migration on an unprecedented scale. Melting glaciers and the expansion of seawater will ensure that sea levels could be as much as 34.5 inches (88 cm) higher by the end of the century, drowning low-lying coastal regions and exacerbating the impact of storms.

Despite the partial success of Kyoto, it is clearly going to be a long, hard, struggle to reduce the greenhouse gas emissions that are responsible for global warming, as long as countries such as the United States do not limit their polluting activities. Given such a lack of political will, it is difficult not to be pessimistic about the sort of future our children and our grandchildren face. Whatever steps we take now and however quickly we take them, temperatures will continue to rise for hundreds of years and sea levels for thousands of years.

There is little chance, therefore, of us avoiding the coming storm, but we can start to prepare for it. Mark Maslin's book helps us do just this, painting a picture of a more hazardous future for all. Perhaps by drawing attention to just how global warming poses a threat to everyone—the rich and the poor, the prince and the pauper—*The Coming Storm* can play a small, but vital, role in shocking the indifferent out of their complacency and saving our temperate world.

Opposite *Tornadoes have generated the fastest winds ever recorded—over 300 miles (483 km) per hour.*

Top *Hurricanes can be more than 1,000 miles (1,600 km) wide, with wind speeds of excess of 200 miles (320 km) per hour. Each year a couple of big hurricanes, such as Floyd in 1999, threaten the east coast of the USA and the Caribbean.*

Bill McGuire

Benfield Greig Professor of Geohazards
University College London

INTRODUCTION

In the twenty-first century, we humans feel like the masters of our Earth: we can send men to the moon and we can build great skyscrapers. We often forget that nature can destroy anything we build. Storms are among the most feared and the most fearful of all natural disasters, with hurtling winds flattening everything in their path, torrential rainfall causing floods that wash away whole towns, and storm surges creating massive ocean waves that smash down, decimating coastal settlements.

Left *Hurricane Floyd hit Florida on September 14, 1999, forcing many evacuations–including that of NASA's Kennedy Space Center.*

Below *Tornadoes are one of the most feared storms as they appear with very little warning, have unpredictable pathways, and are so destructive that nothing can stand in their way.*

Storms, the violent expression of our climate system, are driven by the need to export heat from the hot tropics to the cold poles. Blizzards pound northern Europe and America, monsoons drench Asia and Amazonia, hurricanes race across the tropical oceans, and tornadoes spin wildly over the continents. Nowhere, it seems, is safe from storms.

Of all the storms, the hurricane is king. It takes 500 trillion horsepower to drive hurricane winds–this is the equivalent to one Hiroshima-style atomic bomb exploding every second, or the power to supply the United States with energy for 100 years. Every year hurricanes cause immense damage and loss of life. In the United States, the damage bill from the mainland landfall of hurricanes has averaged US$2 billion per year for the last 50 years. We must thus stand humbled in the face of the extreme power of nature.

STORMY PLANET EARTH

The Coming Storm will investigate the different storms created by our climate system. Over 65 percent of all natural disasters are caused by storms and floods; a major focus of the book is how these storms are caused, exactly what destructive effects they have on humankind, and how storms are influenced by the infamous El Niño.

PREDICTING STORMS

We will see that we are not completely powerless in the face of savage weather. Great steps forward have been made in predicting storms. Meteorologists and climatologists have been able to dramatically improve weather prediction: three-to-four-day forecasts, for example, are now as accurate as the two-day forecasts of 15 years ago. Much progress has been made in evacuation

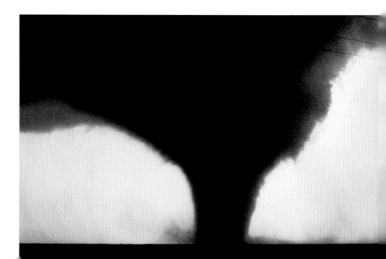

procedures; in addition, education, and relief and rescue work have significantly cut death tolls. However, as we all know, development of this kind costs money. Unfortunately, the majority of the world is unable to afford adequate provision for storm prediction and recovery.

THE FUTURE

In the final section of the book, we gain insight into the science of global warming and its effect. Evidence suggests that we should start planning for more frequent and more powerful storms owing to climate change caused by global warming. The climate system will become less predictable, as we really have no idea what will happen in a warmer world. We know that the amount of carbon dioxide in the atmosphere has been steadily rising for the last hundred years, and if this trend continues, the world could warm up by as much as 9°F (5°C) in the next 100 years.

Top Lightning kills more people each year in the US than either tornadoes or hurricanes. Here, a nighttime multiple cloud-to-ground lightning stroke is captured by means of time-lapse photography.

Right Blizzards and ice storms are a major threat in winter, with driving snow and ice, high winds, and near-zero visibility.

THE AUTHOR'S RESEARCH

Much of my own scientific research has been aimed at understanding the surprises future climate may have in store for us. The two most worrying scenarios, both of which I have worked on, are discussed in *The Coming Storm*:

The first is whether the deep ocean currents will be affected by global warming. Ice melting in either the Arctic or Antarctic could switch off the world's deep-ocean circulation. If it occurs in the Arctic first, it will cause freezing winters in western Europe. If it occurs in the Antarctic first, it could cause the global sea level to rise by 6.5 feet (2 m). This is a scary catch-22 situation, particularly as the possibility of the Antarctic ice shelves collapsing is no long science fiction. In March 2002, the Larsen B ice shelf completely disintegrated in less than 31 days, sending 500 million billion tons of ice into the Southern Ocean.

The second problem is gas hydrates. This mixture of water and methane is sustained as a solid at very low temperatures and very high pressures. If the oceans and the permafrost become warmer, the gas hydrates could break down, pumping huge amounts of methane into the atmosphere. If enough is released, temperatures would rise, releasing more gas hydrates—a run-away greenhouse effect.

THE WAY FORWARD

However, it is not all doom and gloom: we are constantly increasing our understanding of how the global climate system works. This improves our ability to predict weather and storms, and also tells us more about how storms will be affected by global warming. There is even good news on the global warming front: 186 countries (excluding the United States) have agreed to an international treaty on reducing carbon dioxide. The agreement may be small, but it is a step in the right direction—a step which must be the first of many. It's not too late to save our world, but we have to increase the pace and make sure everyone is involved—every individual and every national government. There is no time left for complacency.

WHAT YOU CAN DO

Small changes to your lifestyle could have a big impact on global warming. Here are a few ideas about how you can make the difference—and save yourself money:

Left *Pollution from industry and cars is changing the composition of the atmosphere, causing global warming. The major question is whether global warming will cause there to be more and meaner storms.*

Opposite page *Hurricanes have little mercy, flattening in minutes homes that have taken many generations to build and have been much loved.*

ENERGY

- Turn off televisions, stereos and computers when not in use. Leaving them on standby uses 50 percent of the energy used on full power.
- Buy energy-efficient domestic appliances.
- Turn off lights and use energy-efficient lightbulbs.
- Insulate your home—6 inches (15 cm) of loft insulation can save around 20 percent of heating costs.

WATER

- Don't overfill the kettle when making hot drinks.
- Avoid baths as they use three times as much water as showers.
- Turn off the taps while you brush your teeth.
- Put a full load of washing in the machine, and use the economy cycle.

WASTE

- Recycle and use recycled bottles, cans, paper, plastics, paper, clothes, and electrical appliances.
- Avoid goods that are over-packaged.

TRANSPORT

- Cycle, walk or use public transport rather than private cars.
- Buy fuel-efficient cars.
- Make sure your tires are inflated: you add 1 percent to your fuel bill for every six pounds per square inch (0.4 kg per square cm) a tire is under-inflated.

Top *This satellite picture shows the effect of the prolonged drought on the Amazon Basin in 1997. Although this event was attributed to El Niño, many scientists fear that global warming may make the Amazon Basin too dry to sustain rainforest.*

Above *Global warming seems to be accelerating occurrences of El Niño. In Southern California this will mean more frequent torrential rains and floods. The only people who will gain will be the surfers, who will get increasingly large waves.*

CHAPTER 1

Stormy Planet Earth

Much of our planet is lashed by storms. In the tropics, hurricanes, cyclones, and typhoons pound the coasts, while monsoon rainfall can be measured not in inches or millimeters, but in yards or meters every year. In the more temperate zones, tornadoes ravage communities during summer, while in winter polar storms bring blizzards that combine strong winds, driving snow, ice, hail, and air temperatures as low as −22°F (−12°C).

Why is our planet so bad tempered? What causes storms? This chapter will look at how the climate is generated, and at where the different storms occur and why. An examination of each major type of storm will help us to understand them and the damage they cause.

Above *Storms are essential as they are the main way of transporting heat from the subtropics to the temperate regions. With the use of satellites, we can now understand how they form and move around our planet.*

WHAT CAUSES OUR CLIMATE?

Hot and cold earth

Below *This diagram illustrates how different parts of the Earth receive different amounts of energy from the Sun, making the tropics hotter than the poles.*

The climate of our planet is simple—it is caused by the Earth's equator receiving more heat than the poles.

Above *The dark, dense Amazonian rainforest receives and absorbs huge amounts of energy from the Sun.*

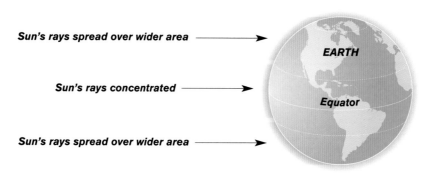

SUN

Sun's rays spread over wider area →

Sun's rays concentrated →

Sun's rays spread over wider area →

EARTH

Equator

POLES APART

If you imagine the Earth as a giant ball, the closest point to the Sun is the middle, or the equator. The equator is where the Sun is most often directly overhead, and it is here that the Earth receives the

most energy. As you move further north or south away from the equator, the surface of the Earth curves away from the Sun. This means that the Sun's energy is spread over a larger area and, thus, causes less warming. The effect is greatest at the poles: for example, during winter in the Northern Hemisphere, the angle of the surface of the Earth in relation to the Sun is so acute that, for one day a year, the Arctic is completely dark and the Antarctic is completely light. Just imagine 24 hours of darkness or sunlight!

Not only do the poles receive less energy than the equator, they also lose more energy back into space. This is because of albedo, or the reflectivity of the surface. The white snow and ice of the Arctic and Antarctic are very reflective, and bounce a lot of the Sun's energy back into space,

while the dark and much less reflective rainforests at the equator absorb much more energy. These two processes working together mean that the tropics are hot and the poles are very cold. Nature hates this sort of energy imbalance, so energy, in the form of heat, is transported by both the atmosphere and the oceans from the equator to both poles, and this is what causes weather. The average weather conditions for a specific region over a period of time are referred to as the climate.

EARTH IN SPACE

Our climate is controlled by three fundamental factors inherent in the relationship between the Earth and the Sun.

First, the Earth is currently slightly closer to the Sun during Northern Hemisphere winter than it is during the summer—91 million

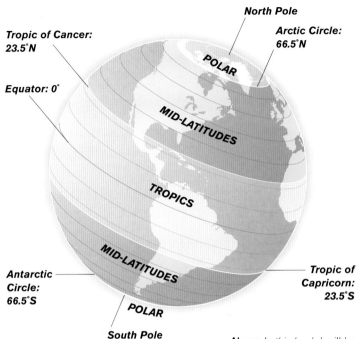

Below *The biggest effect on the Earth's climate is the seasons and it's all to do with the way the Earth is tilted.*

Above *In this book I will be referring to the imaginary lines called latitudes which divide the world into the tropics, the poles, and the area in between called the temperate or mid-latitudes.*

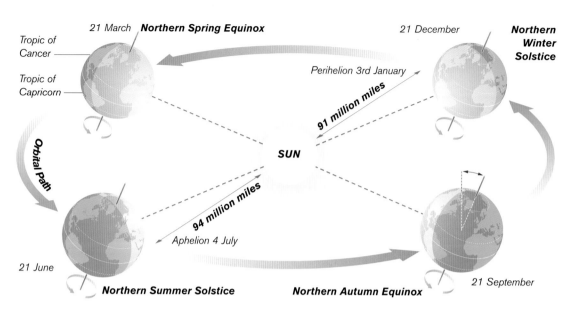

21 March **Northern Spring Equinox**

Tropic of Cancer

Tropic of Capricorn

21 December

Northern Winter Solstice

Perihelion 3rd January

91 million miles

Orbital Path

SUN

94 million miles

Aphelion 4 July

21 June

Northern Summer Solstice

Northern Autumn Equinox

21 September

Left *The four seasons of the Northern Hemisphere relate to the Earth's orbit around the Sun. The Earth's orbit is tilted at approximately 23.5° and thus between the latitudes of 23.5°N (Tropic of Cancer) and 23.5°S (Tropic of Capricorn) the noonday Sun may be directly overhead for at least part of the year.*

Below *This diagram shows how the ocean and atmosphere contribute to heat transport away from the tropics. Note that mid-latitude storms are responsible for much of the atmosphere heat transport between 30° and 60°.*

miles (146 million km) compared with 94 million miles (160 million km). However, we all know that winters aren't warmer than summers! This is because the Earth's tilt, which causes the seasons, has more effect on the climate than the planet's distance from the Sun.

A third factor is the daily rotation of the Earth, which provides us with night and day.

THE SEASONS

The Earth's axis of rotation is tilted at an angle of 23.5°. This angle is fixed in relation to the plane through which the Earth orbits the Sun.

The seasons are caused by variations in the angle of the sunlight hitting the Earth. For half the year each hemisphere leans toward the Sun, and for the other half it leans away from it. If the Earth were straight on its axis, then

we would not have spring, summer, fall, and winter. We would not have the massive change in vegetation in the temperate latitudes and we would probably not have the monsoon or the hurricane seasons in the tropics.

If we take December as an example, the Earth's axis means that the Northern Hemisphere is leaning away from the Sun, so sunlight hits the Northern Hemisphere at a greater angle, spreading its energy over a wider area and keeping temperatures low: it is winter. The lean is so great at midwinter that, in the Arctic, the sunlight cannot reach the surface—so there is one day, 24 hours, of complete darkness. At the same time, there are 24 hours of light in the South, at the Antarctic, where the bottom of the Earth is leaning toward the Sun and the sunlight is more directly overhead.

The seasons cause by far the most dramatic change in our climate. Winter temperatures in New York, for example, can be as low as -4°F (-20°C), while summer temperatures can rise to over 95°F (35°C). Moreover, as we will find out, the seasons are one of the major causes of storms.

NIGHT INTO DAY

The third big factor affecting the climate of the Earth is its daily rotation. First, this takes the Earth in and out of darkness. Depending on the seasons, different areas get varying amounts of daily light. The days can vary from 24 hours' daylight to 24 hours' darkness at

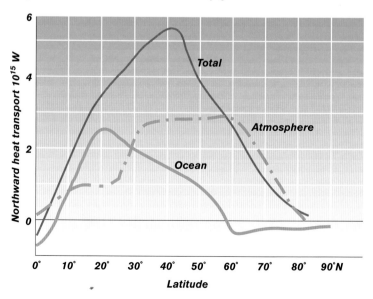

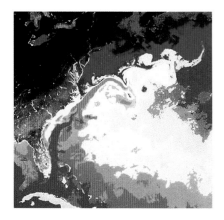

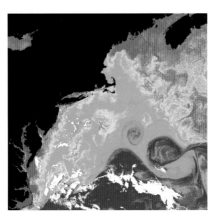

Left *The Gulf Stream, seen here from space, moves an immense amount of warm sea water northward. This helps transport heat away from the tropics to the mid-latitudes. In these satellite pictures, the warmest temperatures are represented by red tones and the coldest temperatures by blue tones.*

the poles, to an average of 12 hours' daylight daily at the equator. This change in the daylight compounds the seasonal contrasts, because not only do you get more direct overhead sunlight during summer, but you also get direct overhead sunlight for longer.

MOVING THAT HEAT AROUND

The spinning of the Earth also makes the transportation of heat away from the equator more complicated. This is because it makes everything else (including the atmosphere and oceans) turn. In the Northern Hemisphere, air and ocean currents are pushed to the right of the direction in which they are traveling, while in the Southern Hemisphere they are pushed to the left.

This deflection is called the Coriolis Force, and its strength increases the closer you get to the poles. An everyday example of this is the way water spirals down a plughole: in the Northern Hemisphere water flows clockwise, while in the Southern Hemisphere it flows counter-clockwise.

But why do the ocean currents and winds deflect in this way? Imagine the currents and winds as a missile fired directly north from the equator. Because the missile is fired from the Earth, which is spinning eastward, the missile also moves east. As the Earth spins, the equator has to move at great speed through space to keep up with the rest of the planet, as it is the widest part. Going north or south away from the equator, the surface of the Earth curves in, so it does not have to move as fast to keep up. So in one day, the equator must move around 24,850 miles (40,074 km—the diameter of the Earth)—a speed of 1,035 miles (1,670 km) per hour, while the Tropic of Cancer (23.5°N) only has to move 22,785 miles (36,750 km)—a speed of 950 miles (1,530 km) per hour. The imaginary missile fired from the equator has the same eastward speed of the equator; as it moves toward the Tropic of Cancer, the surface of the Earth is not moving as fast as the missile. This gives the missile the appearance of moving northeast, since it is moving faster eastward than the area into which it is moving. The closer you get to the poles the greater this difference in speeds, so the greater the deflection to the east.

Right *Different parts of the Earth surface move at different velocities. (**A**) If a missile were fired northward from the equator, it would be moving to an area with a slower velocity. (**B**) It seems to move to the right. (**C**) This relative bending is called the Coriolis Force.*

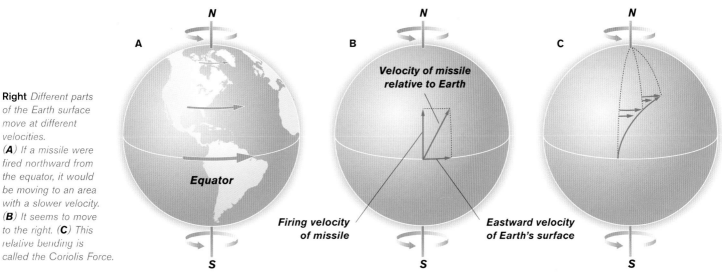

THE ATMOSPHERE
A mixture of gases

The atmosphere begins at the surface of the Earth and extends upward for 80 miles (130 km). In relation to the Earth's size, it is a very thin layer—proportionally less than the thickness of an orange rind.

The layer in which weather takes place is only about 10 miles (16 km) thick. As the oceans also play an important part in controlling our weather and climate, and they are on average about 2.5 miles (4 km) deep, the total thickness of the layer controlling our climate is 12.5 miles (20 km) thick.

The atmosphere is a mixture of gases. Significantly, these gases are mixed in remarkably constant proportions up to about 50 miles (80 km) above the surface of the Earth. Four gases—nitrogen, oxygen, argon, and carbon dioxide—account for 99.98 percent of air by volume. The greenhouse gases (see pages 90–91), despite their relative scarcity, have a large effect on the thermal properties of the atmosphere. They can absorb heat reflected from the Earth, thus warming the atmosphere.

OXYGEN

Oxygen, the gas that sustains all life on Earth, is constantly recycled between the atmosphere and biological processes of plants and animals. It combines with hydrogen to produce water, which in its gaseous state (water vapor) is one of the most important components of the atmosphere.

Oxygen also forms ozone—made up of three oxygen atoms. This is a very important gas, as it forms a thin layer in the upper stratosphere that filters out harmful ultraviolet radiation. A lot of ozone has been destroyed by our use of CFCs (chlorofluorocarbons), and there are holes in the ozone layer over the Arctic and Antarctic. However, the world's governments have now agreed to stop the use of CFCs and related compounds.

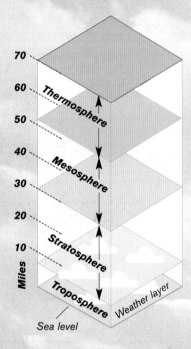

70
60
50
40
30
20
10

Thermosphere

Mesosphere

Stratosphere

Troposphere Weather layer

Miles

Sea level

Above *The atmosphere is divided into four layers, but weather and thus climate only occur in the troposphere, the bottom 8 miles (12.8 km) of the atmosphere.*

Right *Different cloud types form at characteristic heights in the troposphere.*

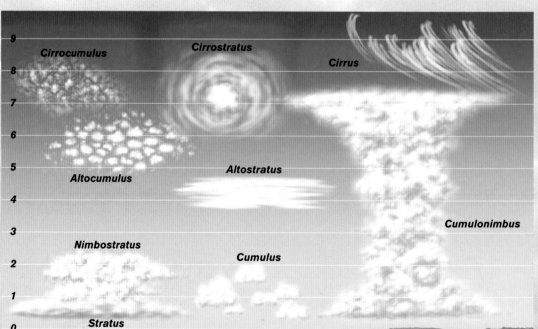

Cirrocumulus

Cirrostratus

Cirrus

Altocumulus

Altostratus

Cumulonimbus

Nimbostratus

Cumulus

Stratus

9
8
7
6
5
4
3
2
1
0

Miles

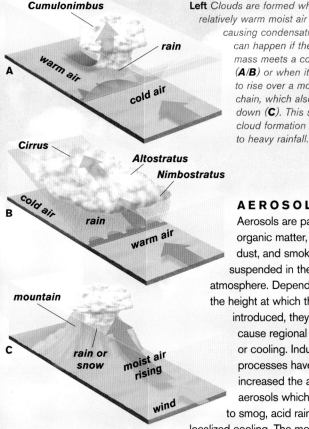

Cumulonimbus

rain

warm air

cold air

A

Cirrus

Altostratus

Nimbostratus

cold air

rain

warm air

B

mountain

rain or snow

moist air rising

wind

C

Left *Clouds are formed where relatively warm moist air is cooled, causing condensation. This can happen if the warm air mass meets a cold air mass (**A/B**) or when it is forced to rise over a mountain chain, which also cools it down (**C**). This sort of cloud formation often leads to heavy rainfall.*

AEROSOLS

Aerosols are particles of organic matter, sea salt, dust, and smoke suspended in the atmosphere. Depending on the height at which they are introduced, they can cause regional warming or cooling. Industrial processes have increased the amount of aerosols which has led to smog, acid rain, and localized cooling. The most important effect of aerosols is the formation of clouds. Without these minute particles, water vapor cannot form clouds.

WATER VAPOR

The most important greenhouse gas–water vapor–makes up 0–2 percent by volume of the atmosphere. Water vapor is essential for the formation of clouds. As water changes from a gas to a liquid, it releases energy, and it is this energy which can fuel storms as large as hurricanes.

CARBON DIOXIDE

Carbon dioxide is a major greenhouse gas, and in small quantities it is very important for warming the Earth. Usually the amount of carbon dioxide in the atmosphere is balanced–plants use it up in photosynthesis, and plants and animals produce it in respiration. However, human industry over the last century has been pumping a lot of carbon dioxide into the atmosphere. As we will see later in the book, this does not have a direct effect on weather, but it contributes to global warming which may change our climate immeasurably.

STRUCTURE OF THE ATMOSPHERE

The atmosphere can be divided into a number of clearly defined layers, mainly based on temperature (see below).

Troposphere

This is the lowest layer of the atmosphere, where atmospheric turbulence and weather are at their most marked. The troposphere contains 75 percent of the total molecular

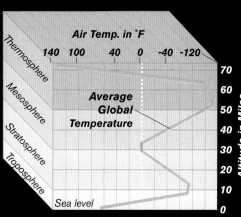

Air Temp. in °F

Thermosphere

140 100 40 0 -40 -120

Mesosphere

Average Global Temperature

Stratosphere

Troposphere

Sea level

Altitude in Miles

70
60
50
40
30
20
10
0

mass of the atmosphere and virtually all the water vapor. Throughout this layer, there is a decrease in temperature at a mean rate of about 44°F (24°C) roughly every half mile (km). The whole zone is capped by a layer known as the tropopause, which acts as a lid of the troposphere and on weather; the temperature begins to rise again at this point.

Stratosphere

The second major atmospheric layer extends upward from the tropopause to about 31 miles (50 km) above sea level. Although the stratosphere contains much of the ozone, the maximum temperature caused by the absorption of ultraviolet radiation occurs at the "stratopause," where temperatures may exceed 32°F (0°C). This large temperature increase is due to the relatively low density of the air at this height.

Mesosphere

Above the stratopause, average temperatures decrease to a minimum of -130°F (-90°C). Above 50 miles (80 km), temperatures begin rising again because of absorption of radiation by both ozone and oxygen molecules. This temperature inversion is called the mesopause. Pressure is extremely low in the mesosphere, decreasing from 1 mb at 31 miles (50 km) to 0.01 mb at 56 miles (90 km). Pressure at the Earth's surface is about 1000 millibar.

Thermosphere

Above the mesopause, atmospheric densities are very low. Temperatures rise throughout this zone due to the absorption of solar radiation by molecular and atomic oxygen.

HADLEY CELLS

Storm zones caused by wind

Named after the eighteenth century scientist, George Hadley. He suggested the Hadley Cell is a pattern of atmospheric circulation, whereby warm air near the equator rises, then cools and sinks as it travels to the poles. Scientists now realize that there are not one but three Hadley Cells.

TRADE WINDS

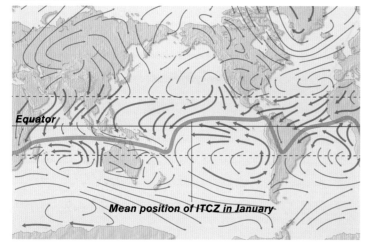

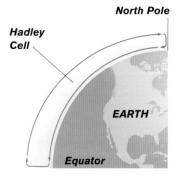

Above and right *Hadley's original model had only one cell with air rising at the equator and sinking at the poles, but later scientists found there were three main cells.*

At the equator, the intense heat from the Sun heats up the air near the surface and causes it to rise high into the atmosphere (remember, hot air rises and cold air sinks). This upward rush of air creates a space, represented by low atmospheric pressure; air is sucked in to this space, producing strong winds, known as the Trade Winds, in both the Northern and Southern Hemispheres.

This causes a problem: the climate system is desperately trying to export heat away from the region around the equator, and these in-blowing winds do nothing

OCEAN CURRENTS

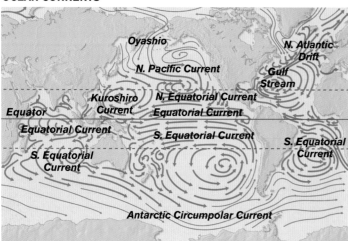

Above left *Surface winds in Northern Hemisphere winter are like the patterns predicted by the Hadley Cell model. However, the continents heat and cool quicker than the oceans, distorting the wind direction. Winds change direction as they move across the equator due to the Coriolis Force.*

Left *The movement of the surface ocean controls the surface winds– this is why they look similar. The western boundary currents transport heat away from the tropics.*

to help with heat removal. So in the tropics, it is the surface currents of the ocean that transport most of the heat. These currents include the Gulf Stream, which takes heat from the Tropical Atlantic Ocean and transports it northward, keeping Europe's weather mild year-round, the Kuroshiro current in the Western North Pacific Ocean, the Brazilian current in the Western South Atlantic Ocean, and finally the East Australian current in the Western South Pacific Ocean.

RISING AIR KICKS UP A STORM

The hot air that has risen high into the atmosphere in the tropics slowly cools due to both its rise and its movement toward the poles. At about 30° latitude it sinks again, forming the subtropical

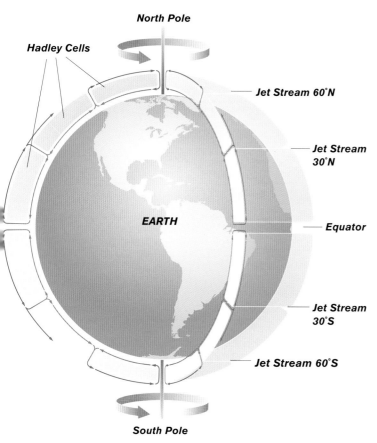

North Pole

Hadley Cells

Jet Stream 60°N

Jet Stream 30°N

EARTH

Equator

Jet Stream 30°S

Jet Stream 60°S

South Pole

rising air drifts over the poles, where it is chilled and sinks, forming out-blowing winds and completing the third Hadley Cell.

Though the general wind patterns of our planet follow this simple three-celled model, in reality they are far more complicated. First, the Earth is spinning, which

adds the influence of the Coriolis Force. Second, friction between the surface of the Earth and the winds means surface winds are influenced by both the shape and the type of surface over which they are trying to flow.

highs. As this air reaches the surface it spreads out, moving both north and south. The southward air links into the first Hadley Cell and becomes part of the Trade Winds system. The northward air forms the Westerlies; from here northward, the atmosphere takes over from the oceans as the major transporter of heat.

It is well known by climatologists that, between 30° and 70° latitude, mid-latitude storms transport most of the heat. The movement of warm subtropical air northward is stopped only when it meets the cold air mass at the Polar Front. The intense cold at the

poles causes air to become super-chilled and sink, causing out-blowing winds.

When this cold polar air meets the warm moist Westerlies, the clash causes the Westerlies to lose a lot of moisture in the form of rain. It also forces the warm subtropical air to rise, as the cold polar air is much heavier. This rising air completes the other two Hadley Cells because, as the air rises, it spreads out to both the north and the south. To the south, this rising air meets tropical air coming northward and sinks, forming the middle Hadley Cell. The northward component of this

HADLEY CELLS

The Hadley Cells explain why there are three main storm zones:
A The Polar Front, where cold dry air meets warm moist air. This is why, when the Polar Front passes over Britain, there is so much rain.
B The Subtropical Highs and the Trade Winds belt—the spawning ground for hurricanes.
C The Intertropical Convergence Zone (ITCZ), where rapidly rising air cools and produces tropical thunderstorms and heavy rain.

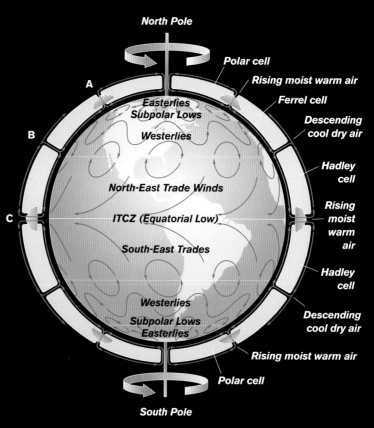

STORM AND CLIMATE ZONES

Our diverse world

The world can be divided up into different climate zones which are controlled by the Hadley Cells and the position of land with respect to mountains and the ocean. We also know those areas of the world that are prone to storms. As you can see, over 80 percent of the world's land mass is affected by some sort of storm.

TORNADOES

A tornado is an extremely tight rotating column of air, which can be observed at a distance as a cone-shaped cloud formation. Tornadoes are among the most violent storms in the world. They are most numerous and devastating in central, eastern, and northeastern United States, where an average of five are reported every day in May. They can also occur in Australia (15 per year), Great Britain, Italy, Japan, and Central Asia.

MONSOONS

Monsoons are massive seasonal rainstorms which occur when air masses from the Northern and Southern Hemispheres clash at the Intertropical Convergence Zone (ITCZ). The Monsoon Belt stretches around the equator and affects tropical South America, South-East Asia, and parts of Central Africa.

Arctic Circle: 66.5°N

Tropic of Cancer: 23.5°N

Equator: 0°

Tropic of Capricorn: 23.5°S

Antarctic Circle: 66.5°S

KEY

	Source area of tropical cyclones		Monsoon rains	→	Hurricane track
					Cyclone track
	Tornado high-risk areas		Limits of polar front storms (any time of year)		Willy-willies (tornadoes)
				→	Typhoon track

HURRICANES

A hurricane is a severe cyclonic tropical storm in which the sustained wind speed must exceed 75 miles (120 km) per hour. The areas in which this type of storm occur are shown on the map. These storms are called hurricanes in the North Atlantic Ocean, Caribbean Sea, Gulf of Mexico, west coast of Mexico, and North-East Pacific. They are called typhoons in the western Pacific Ocean, and cyclones in the Indian Ocean and around Australia. In this book we will refer to such storms as hurricanes.

WINTER STORMS DUE TO POLAR FRONT

The Polar Front is the point at which warm, wet air from the tropics meets cold, dry air from the poles. Where they meet there are strong storms and lots of rain and snow. The Polar Front moves north and south with the seasons, but the worst storms occur in winter. The general areas affected by these Polar Front storms are shown on the map.

CLIMATE ZONES

The climate of the world can be divided into many different types. Climate zones are defined both by the yearly and the seasonal average temperature and rainfall. (Seasonality must be taken into account, as many deserts have the same annual rainfall as London; however, the rain falls over a very short period of time, so the rest of the year is extremely arid.) Another indicator of the climate zones is the major vegetation types.

Forest

MOUNTAIN RANGES

Mountain regions have their own climate. Because as air is forced to rise over them, it is cooled down. This means that it cannot hold as much water vapor, hence mountansides receive a lot of rainfall. Combined with the complicated geography of the mountain regions, storms can be easily produced and can form rapidly, trapping the unwary climber or walker in life-threatening conditions.

KEY

▨ *Mountain belts of Cenozoic age*	▨ *Escarpments*
▨ *Partially eroded mountain belts of Mesozoic and late Palaeozoic age*	▨ *Rift valleys*

Steppe

Mountains

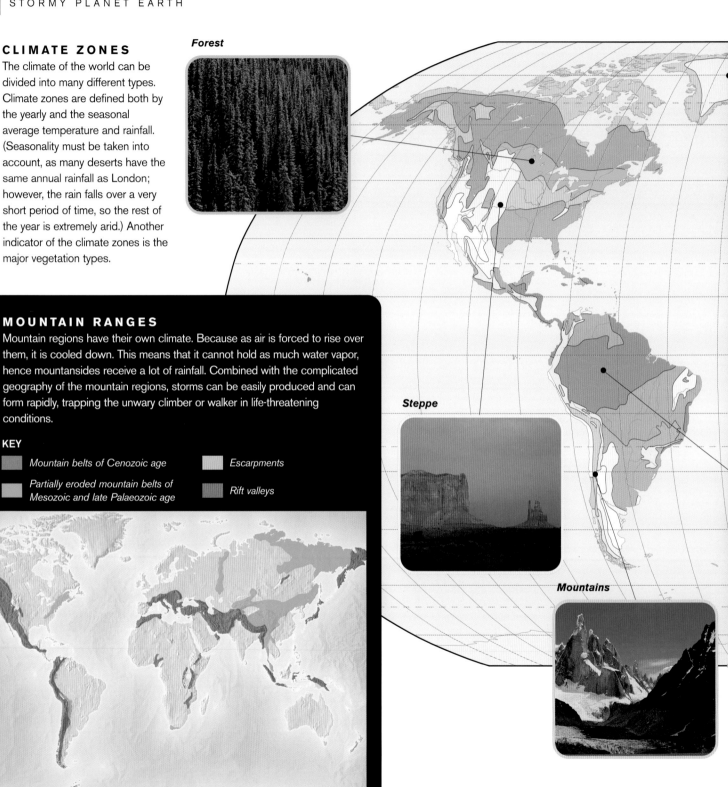

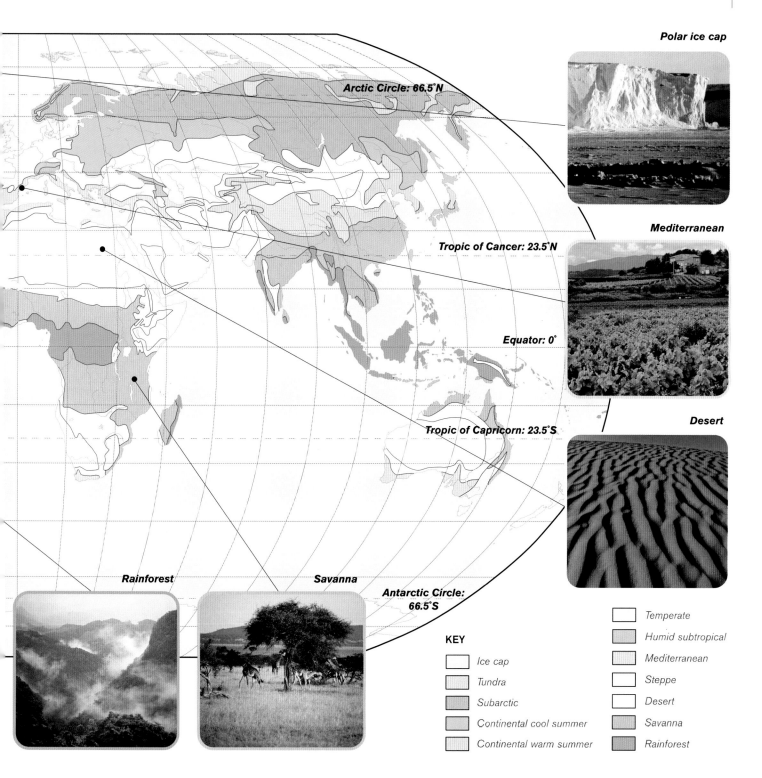

Polar ice cap

Mediterranean

Desert

Arctic Circle: 66.5°N

Tropic of Cancer: 23.5°N

Equator: 0°

Tropic of Capricorn: 23.5°S

Antarctic Circle: 66.5°S

Rainforest

Savanna

KEY

☐	Ice cap
☐	Tundra
☐	Subarctic
☐	Continental cool summer
☐	Continental warm summer

☐	Temperate
☐	Humid subtropical
☐	Mediterranean
☐	Steppe
☐	Desert
☐	Savanna
☐	Rainforest

HURRICANES

Tropical storms run amok

A rotating mass of thunderstorms, a roaring vortex of wind up to 1,000 miles (1,600 km) wide, hurricanes wreak havoc on large areas of the world on a regular basis. But exactly what is a hurricane—and what causes them?

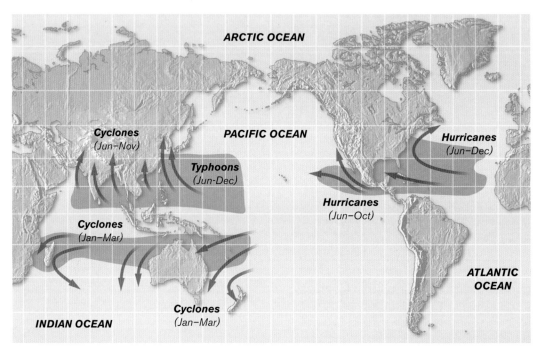

For a storm to be officially classified as a hurricane, the sustained wind speed must exceed 75 miles (120 km) an hour—although a fully developed hurricane can sustain wind speeds of more than125 miles (200 km) an hour. In the Western Pacific Ocean such storms are called typhoons; in the Indian Ocean and Australia, they are known as tropical cyclones.

WHAT CAUSES A HURRICANE?

Born in the warm waters of the North Atlantic Ocean, Caribbean Sea, Gulf of Mexico, west coast of Mexico, and North-East Pacific Ocean, hurricanes occur in the tropics between 30°N and 30°S (but not near the equator, as there is not enough atmospheric variation to generate them).

KEY

 Typical paths of intense tropical storms

Tropical storm breeding grounds

THE SAFFIR-SIMPSON HURRICANE DAMAGE SCALE			
Category	Damage	Winds–mph (kph)	Storm surge–ft (cm)
1	minimal	74–95 (119–153)	4–5 (122–181)
2	moderate	96–110 (154–177)	6–8 (182–273)
3	extensive	111–130 (178–209)	9–12 (274–395)
4	extreme	131–155 (210–249)	13–18 (396–548)
5	catastrophic	155+ (249+)	18+ (549+)

HURRICANE-PRONE REGIONS OF THE WORLD

Below *Hurricane winds are so strong, sometimes gusting at over 200 miles (more than 320 km) per hour that even the strongest trees cannot withstand them.*

Opposite top *Hurricanes are truly massive storms, ranging in size from 100 miles (160 km) to over 1,000 miles (1600 km) across.*

Opposite bottom *Hurricane Andrew in 1992 caused US$30 billion of damage and is the costliest natural disaster in U.S. history.*

The intense heat of the sun close to the equator warms the land, which in turns heats the air. This hot air rises and sucks air from both the north and south, producing the Trade Winds. As the seasons change, so does the point at which the Trade Winds clash—the Intertropical Convergence Zone (ITCZ). It is only within this zone that the temperature and humidity are just right to produce a hurricane.

To generate a hurricane, the sea temperature must be above 79°F (26°C) for at least 200 feet (60 m) below the surface, and the air humidity must be about 75–80 percent. This combination of heat and water vapor sustains the storm once it has started.

Initially, the warm ocean heats the air above, causing it to rise. This produces a low-pressure area that sucks in air from the surrounding area. As water evaporates quickly from the hot surface of the ocean, the rising air contains great amounts of water vapor. As the air rises and cools, it can no longer hold as much water vapor, resulting in condensation and cloud formation. This transformation from water vapor to water droplets releases energy called "latent heat," which in turn warms the air even more, causing it to rise even higher. This feedback can make the air within a hurricane rise to over 33,000 feet (10,000 m) above the ocean. The "eye" of the storm is where all this action takes place, and the spiraling, rising air it produces creates a huge column of cumulonimbus clouds.

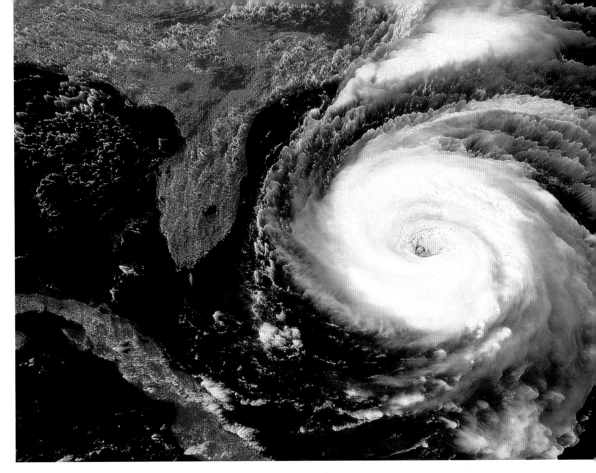

TYPHOON ALLEY

An average of 31 tropical storms roam the western North Pacific every year, with typhoons smashing into South-East Asia from June to December. The countries most at risk are Indonesia, Hong Kong, China, and Japan. This area, known as Typhoon Alley, is plagued with so many typhoons because of the warm pool of ocean water that sits in the western tropical Pacific Ocean. Trade Winds and the ocean current push the surface water, warmed by the tropical sun, to the far western side of the North Pacific Ocean, and hence the waters of Typhoon Alley are always at the right temperature for hurricanes to be formed.

KETTLE CUMULONIMBUS

To see a mini version of the process that forms a cumulonimbus, watch what happens as steam comes out of a boiling kettle. Hot air rises from the lip of the kettle and then hits the colder air outside, forming steam—in effect, a mini-cloud. The steam's heat is the result of all the energy being released as the water vapor changes from a gas back to a liquid.

When the air inside the hurricane reaches its highest level, it flows outward from the eye and produces a broad canopy of cirrus cloud. The air then cools and falls back to sea level, where it is sucked back into the center of the storm. Because of the Coriolis Force—the deflection caused by the rotation of the earth—the air that is sucked into the bottom of the hurricane spins into the storm in a counterclockwise direction, while the air escaping at the top spins out in a clockwise direction. (The Coriolis Force is what makes bathwater go down the drain clockwise in the Northern Hemisphere and counterclockwise in the Southern Hemisphere.) As a hurricane moves over land it tends to lose force because, unlike a temperate storm, it is driven by the latent heat of the water's condensation.

THE ANATOMY OF A HURRICANE

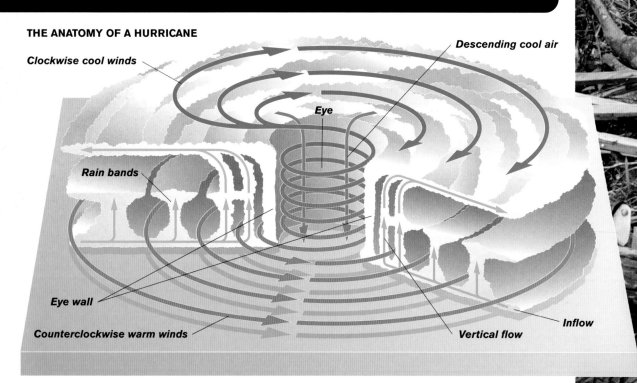

Clockwise cool winds

Descending cool air

Eye

Rain bands

Eye wall

Counterclockwise warm winds

Vertical flow

Inflow

Right *Hurricanes are so strong that they can flatten houses and snap trees as if they were matches.*

Opposite top *Hurricane Hugo in 1989 damaged many areas along the U.S. Eastern seaboard, including this recently developed condominium in Virginia. Most of the damage was caused by the airborne debris thrown around by the massive winds.*

Opposite bottom *Storm surges and floods due to heavy rains are also a major hazard caused by hurricanes. In this example, a town in Fairfield, New Jersey, is flooded by the swollen Passaic River.*

HURRICANE FACTS

- Hurricanes can vary in size from 60 miles (100 km) to over 950 miles (1500 km) across.
- Hurricanes can form gradually over a few days or rapidly in the space of six to 12 hours. Typically, the hurricane stage will last two to three days, taking about four to five days to die out.
- Only ten percent of centers of falling pressure over the tropical oceans give rise to fully fledged hurricanes.
- In a busy year, up to 50 tropical storms might develop to hurricane levels.
- Sometimes hurricanes can form outside of the tropics, as evidenced in Britain in 1987.
- It takes 500 trillion horsepower to whirl hurricane winds at the tremendous speed of over 200 miles (322 km) per hour—equivalent in size and power to the atomic bomb that destroyed Hiroshima exploding every second, or enough power to meet the United States' energy needs for a hundred years.

TORNADOES
Terrifying twists of nature

Whirling through the air at speeds in excess of 250 miles (400 km) per hour, a tornado is one of the most destructive storms on Earth.

Tornadoes are most numerous and devastating in the Central, Eastern, and North-Eastern United States, where an average of five per day are reported every May. They also occur in Australia (15 per year), the United Kingdom, Italy, Japan, and Central Asia. Most fatalities occur in the United States: between 1950 and 1978, 689 tornadoes were classified as "killers."

WHAT CAUSES TORNADOES?

The formation of a tornado is precipitated by warm, moist air near the ground meeting cold, dry air above—so, although tornadoes can form over tropical oceans, they occur more commonly over land. When the ground is heated by the Sun, warm moist air rises and cools, forming massive cumulonimbus

Above *This is the oldest known photograph of a tornado, taken in 1884 22 miles (35 km) southwest of Hazard in South Dakota.*

Left *Tornado seen spiraling through Miami.*

clouds. The strength of the updraft determines how much of the surrounding air is sucked into the bottom of the tornado.

Two things help the tornado to rotate violently. The first is the Coriolis Force (see pages 16 and 17), and the second is the high-level jet stream. As it passes over the top of the storm, the jet stream catches one edge and adds an extra twist to the rotating center. The conditions in which they are formed means tornadoes can often occur beneath thunderstorms and hurricanes.

MEASURING TORNADO STRENGTH

Wind-measuring instruments are destroyed by the high speeds of tornadoes (some are in excess of 250 miles (400 km) per hour). According to most

THE ANATOMY OF A TORNADO

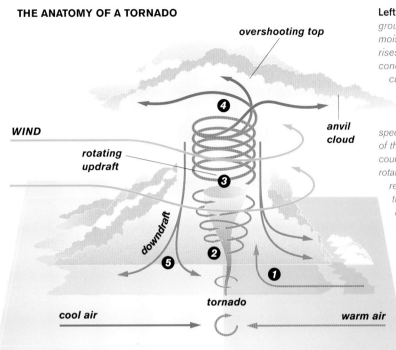

overshooting top

WIND

anvil cloud

rotating updraft

downdraft

tornado

cool air

warm air

Left (*1*) *Intense heating of the ground in summer warms the already moist air, causing it to rise.* (*2*) *As it rises it cools and the water vapor condenses, forming a towering cumulonimbus cloud and releasing heat that sustains the updraft.* (*3*) *The updraft meets winds of different directions and speeds. This, together with the rotation of the earth, causes it to spiral counterclockwise, creating a vortex (a rotating column of air).* (*4*) *On reaching the warmer stratosphere, the overshooting top of the updraft stalls, as it is cooler than its surroundings. Winds from the jet stream blow the top of the storm across the bottom of the stratosphere, giving it an extra spin and creating the characteristic anvil cloud.* (*5*) *Downdrafts of cool air descend through the storm, exiting at its base.*

Below *These four photographs taken in 1966 in Enid, Oklahoma, show the formation and development of a mature and very dangerous tornado.* (*1*) *Early stages of the formation of the tornado.* (*2*) *The tornado continues to develop and dust and debris can be clearly seen carried half way up to the cloud base.* (*3*) *The tornado in its mature phase.* (*4*) *The tornado has reached its maximum size and strength and is extremely powerful and dangerous.*

4

THE FUJITA SCALE	
F$_1$	Small-scale tornado, usually (but not always) harmless
F$_2$	Capable of pushing vehicles off roads and tearing the roofs off houses
F$_3$	Severe structural damage to wooden houses, vehicles lifted off the ground
F$_4$	Brick buildings demolished, vehicles picked up and carried for over a mile (2 km)
F$_5$	Terrifying wind strength, usually maintained for only a few seconds
F$_6$	Theoretically possible, but never been observed

THE PEARSON SCALE measures the length and width of the tornado path.

Pl	0.3–1 mile (0.5–1.5 km)	Pw$_0$	18–50 ft (5.5–15.5 m)
Pl$_1$	1–3 miles (1.5–5 km)	Pw$_1$	50–164 ft (15.5–50 m)
Pl$_2$	3–10 miles (5–16 km)	Pw$_2$	164–525 ft (50–160 m)
Pl$_3$	10–31 miles (16–50 km)	Pw$_3$	525–1640 ft (160–500 m)
Pl$_4$	31–100 miles (50–160 km)	Pw$_4$	1640–4920 ft (500–1,500 m)
Pl$_5$	100+ miles (160+ km)	Pw$_5$	4920–16404 ft (1,500–5,000 m)

Using these two scales, the potential damage and the area likely to be affected can be classified and predicted.

scientists, the top wind speed in the strongest tornadoes is about 280 miles (450 km) per hour.

Tornado strength is measured using two different scales (see page 31). The Fujita Scale (F) classifies the strength of a tornado based on the speed at which it rotates (the data provided by satellite). Pearson Scale measures the length and width of the tornado path.

TORNADO ALLEY

Nowhere else in the world sees weather conditions so perfect for the formation of tornadoes as Tornado Alley in the United States. The area includes central Texas, Oklahoma, and Kansas in the peak season, but it can also include Nebraska and Iowa: the area expands through spring and summer as heating from the Sun grows warmer and the flow of warm, moist air from the Gulf of Mexico spreads farther north. The frequency of tornadoes in this area is a result of several factors.

■ Beginning in spring and continuing through summer, low-level winds from the south and southeast bring a plentiful supply of warm tropical moisture up from the Gulf of Mexico into the Great Plains.

■ From the eastern slopes of the Rocky Mountains or from out of the deserts of Northern Mexico come other flows of very dry air that travel about 3,000 feet (900 m) above ground.

■ From 10,000 feet (3,000 m) the prevailing westerly winds, sometimes accompanied by a powerful jet stream, race overhead, carrying cool air from the Pacific Ocean.

Above *As this particular tornado passes the farmhouse, the winds are so strong that the suction alone is enough to topple the barn.*

Right *This tree was lucky: the tornado only passed close enough to rip off the top branches. Any closer and it would have uprooted the whole tree.*

TORNADO FACTS

- Other storms similar in nature to tornadoes are whirlwinds, dust-devils (weaker cousins of tornadoes occurring in dry climates), and waterspouts (tornadoes that occur over water).
- In the United States, nearly 90 percent of tornadoes travel from the south-west to the north-east, although some follow zigzag paths.
- Weak or decaying tornadoes have a thin, ropelike appearance, while the most violent tornadoes have a broad, dark, funnel shape that extends from the dark wall of the thunderstorm's cloud.
- On average, tornadoes travel at 35 miles (60 km) per hour—though some have been clocked at over 70 miles (113 km) per hour.
- Most tornadoes occur between 3 and 9 pm.
- They usually only last about 15 minutes, spending only a matter of seconds in any single place. However, on March 18, 1925 a single tornado traveled 219 miles (350 km) in 3 hours through Missouri, Illinois, and Indiana. Six hundred and ninety-five people were killed.

Above *Areas over which a tornado pass can be completely flattened.*

TORNADO ALLEY, U.S.A.

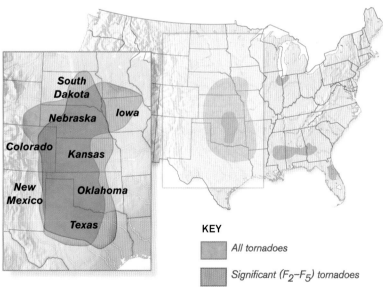

KEY

All tornadoes

Significant (F₂–F₅) tornadoes

Above *Tornado Alley in the United States—this area has ideal weather conditions for the formation of tornadoes.*

POLAR-FRONT STORMS

Clash of the titans

As cold polar air meets the warm air of the subtropics, clouds form—and sometimes fierce storms result.

Background *Large banks of gray and black cloud roll in toward the west coast of Ireland as a depression approaches.*

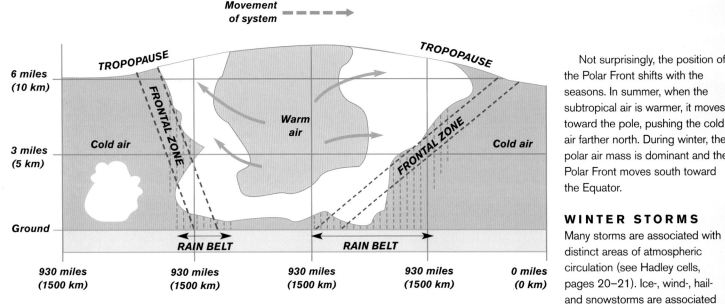

Movement of system

6 miles (10 km)

3 miles (5 km)

Ground

TROPOPAUSE

FRONTAL ZONE

TROPOPAUSE

FRONTAL ZONE

Cold air

Warm air

Cold air

RAIN BELT

RAIN BELT

| 930 miles (1500 km) | 930 miles (1500 km) | 930 miles (1500 km) | 930 miles (1500 km) | 0 miles (0 km) |

Above *Cross-section through the air masses in a depression. This illustration also shows the changes in weather that we experience as a depression passes overhead—from cold and dry to wet and relatively warm, and finally back to cold and dry again.*

From sudden showers to unexpected bursts of sunshine, often in the course of the same day, the United Kingdom is well known for its changeable weather. The reason for it is the U.K.'s midlatitude climate, which is dominated by the Polar Front—the point at which cold polar air moving southward and warm subtropical air trying to move northward meet. Since warm air holds more water vapor than cold air, clouds form when the two clash, producing rain.

The shape and conditions of the Polar Front are controlled by the upper atmosphere; here, fast "jet streams" push the Polar Front creating a mass of air waves that moves gradually around the Earth, causing variations in the weather.

Not surprisingly, the position of the Polar Front shifts with the seasons. In summer, when the subtropical air is warmer, it moves toward the pole, pushing the cold air farther north. During winter, the polar air mass is dominant and the Polar Front moves south toward the Equator.

WINTER STORMS

Many storms are associated with distinct areas of atmospheric circulation (see Hadley cells, pages 20–21). Ice-, wind-, hail- and snowstorms are associated with either the Polar Front or with high mountain regions and are more extreme during winter.

In the Northern Hemisphere these storms are common over North America, Europe, Asia, and Japan. At the Polar Front, warm air from the subtropics pushing toward the North-East meets cold polar air trying to move toward the South-West. As they try to move past each other, the Coriolis Force

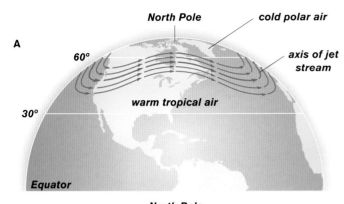

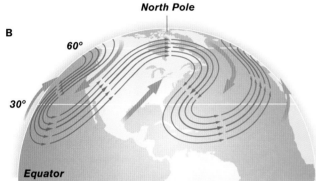

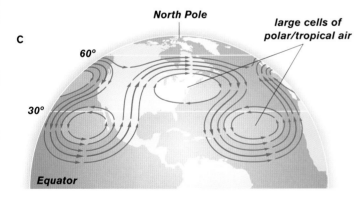

Above *The stages in the development of waves in the northern polar jet stream which flows along the polar limit of the upper westerlies near the tropopause at heights of just over 6 miles (10 km).*

A *The jet stream begins to undulate*
B *Waves become more extreme*
C *Large cells of polar and tropical air become isolated; cyclones and anticyclone storms form along these boundaries*

causes them to spin, resulting in both a warm front and a cold front. At these fronts, the warm moist air rises and cools, forming clouds. Whether those clouds give rise to rain, snow, or hail depends on the temperature difference between the cold and warm air masses.

PICTURE THIS...

Imagine a polar-front wave passing over a town during a period of 24 hours. To begin with, the temperature is relatively cold and the skies are clear. As the warm front approaches, the temperature rises and it starts to lightly rain. The center of the warm-air front then passes overhead and it becomes cloudy and muggy and the rain stops. The cold front follows, causing a drop in temperature and a short period of very intense rainfall. Finally, it's back to cold, clear weather until the next wave passes over.

Storm seen at a distance over the ocean

Snow

For snow to reach the ground, the temperature of the air between the base of the cloud and the ground must be below 39°F (4°C). Otherwise the snowflakes melt as they travel through the air.

This snow crystal shows characteristic hexagonal symmetry. In order to assume a stable crystal arrangement, the snowflake must be arranged as a 6-branched figure.

Hail

For hailstones to form, the top of the storm must be very cold. In the reduced pressure high up in the atmosphere, water droplets can be cooled to less than

Hail stones can vary in size between than 0.2 inch (2 mm) and 8 inches (20 cm,

32°F (0°C), which then collide in the atmosphere and form ice balls or hailstones. If you cut open a hailstone, you will see that the layers of ice that have built up resemble the layers of an onion. The size of a hailstone depends on the strength of the updraft of air, which determines how long i stays in the atmosphere before dropping out.

Blizzards

Blizzards represent the worst winter storm conditions. They combine strong winds, driving snow, ice, and hail, with air temperatures as low as 10°F (−12°C) and visibility of less than 500 feet (150 m).

Top left *Polar-front winter storms can reach a long way south. Here, a Citroën 2CV is covered by snow in the town of Nantes in the west of France.*

Left *Snow can accumulate very quickly. In many areas, such as Quebec, houses need to be dug out from under the snow after a fierce night storm.*

CAUGHT IN THE COLD

In extreme cold conditions the body's core temperature can drop, which can result in death. You may be in danger if:

■ you can't stop shivering
■ your hands shake and your gestures are fumbled
■ your speech is slow and slurred, or even incoherent
■ you stumble and lurch as you walk
■ you are drowsy and exhausted and feel the need to lie down even when outside
■ you have rested but cannot get up

Anyone affected in this way needs a warm hot water bottle, a heating pad, or warm towels on their body; dry clothes and a warm bed; and warm, non-alcoholic, and non-caffeinated drinks. It is important not to massage or rub the individual, as this again takes away heat from the body core where it is most required. Seek medical attention in extreme cases.

Background *Blizzards are the worst winter storms, with near-zero visibility and freezing rain, and it is easy to become trapped. Here a driver tries to make it through a blizzard in Muir of Ord in the Scottish Highlands.*

MONSOONS

Devastating tropical rainstorms

Every year, massive rainstorms occur in countries along the "Monsoon Belt" of the Indian Ocean and South Atlantic, bringing much-needed water for crops—but also flooding, damage, and famine.

INTERTROPICAL CONVERGENCE ZONE (ITCZ)

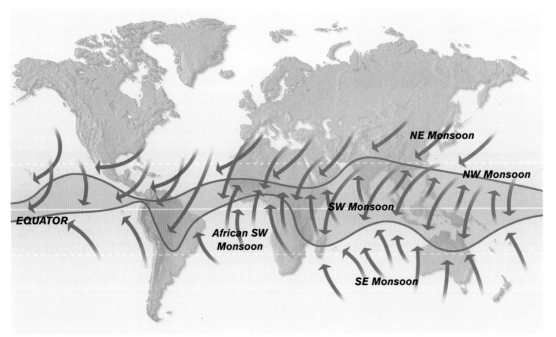

Above *The position of the ITCZ is controlled both by the seasons and by the positions of the continents and oceans. So the simple theoretical ITCZ becomes an undulating area across the globe.*

Right *Monsoons bring essential water for life—but often they bring too much, as can be seen here at Basseterre in St Kitts.*

KEY

⟶ *Dominant winds in July*

⟶ *Dominant winds in January*

⌒ *JULY mean position of ITCZ*

⌒ *JANUARY mean position of ITCZ*

Most of the rains that fall in South-East Asia occur during the summer, and the word "monsoon" comes from the Arabic *mausim*, which means "season." In the tropics, the Sun's energy is most intense when it is overhead and this heats up the land or sea, warming the air above. This warm,

moist air rises, leaving an area of low pressure beneath it into which air from the surrounding area is sucked and this results in the Trade Winds that come from both the Northern and Southern hemispheres. This area is known as the Intertropical Convergence Zone (ITCZ), and the air within it rises and forms huge towering clouds that produce heavy rainfall.

MOVEMENT OF THE ITCZ

The ITCZ moves north and south with the seasons as the position of the most intense sunlight shifts up and down across the equator. The position of the continents is a strong influence, because land heats up faster than the ocean, pulling the ITCZ even further north or south during the respective seasons. An example of this is the Asian summer monsoon, which occurs during the summer in the Himalayas when the lowlands of India heat up. This pulls the ITCZ across the equator into Asia. Because the Southern Hemisphere winds have been

NORTHERN HEMISPHERE: WINTER

Warm humid air

Cold dry air

❸

❷

❶

Moisture from ocean

LOW (ITCZ)

HIGH

Equator

Continent cooler than ocean

NORTHERN HEMISPHERE: SUMMER

Cold dry air

Warm humid air

❹

HIGH

Moisture from ocean

LOW (ITCZ)

Equator

Continent warmer than ocean

Left *The changing conditions in the Northern Hemisphere over both land and ocean in a monsoon climate.*

WINTER
*(**1**) Dry cool air subsides over the continent which is a region of high surface pressure.*
*(**2**) Winds blow off the continent, picking up moisture from the much warmer oceans.*
*(**3**) Moist air rises at the Intertropical Convergence Zone causing abundant rain*

SUMMER
*(**4**) The land is now much warmer than the ocean and so it heats up the air, causing it to rise. This produces a region of low pressure which moves the ITCZ and its rain zone northward toward the interior of the continent.*

Above *Monsoons are caused by the shifts in the Intertropical Convergence Zone (ITCZ). This can be seen over the ocean as an area of towering clouds.*

pulled across the warm Indian Ocean, they are warm and moist, so when they are forced to rise and cool over India they produce heavy rains throughout South-East Asia and as far north as Japan.

During the Northern Hemisphere winter, the ITCZ moves south of the equator, which means that in South-East Asia warm, moist winds from the northern Tropical Pacific are also dragged southward across the continent into the Southern Hemisphere. As a result, areas such as Indonesia and Southern China get two monsoon periods each year, making this the most fertile region on Earth, supporting more than two-fifths of the world's population.

Right *The Amazon River, the mightiest river in the world, carrying 20% of all fresh water transported to the oceans.*

AMAZON MONSOONS

During summer in the Southern Hemisphere, the continent of South America heats up. Rising air leaves an area of low pressure at ground level that acts like a vacuum and sucks in the surrounding air. This pulls the convergence zone between the northern and southern tropical air southward over Brazil. The southward shift of the ITCZ brings heavy rain, since the air pulled across the Equator from the north originates over the warm, tropical Atlantic Ocean. This produces the Amazon Monsoon, which feeds the mighty Amazon River and the greatest extent of rain forest on the planet.

The Amazon Basin covers an amazing 2.7 million square miles and the Amazon River delivers 20 percent of all the fresh water entering the world's oceans. Without the monsoon rains, the most diverse habitat in the world would not exist.

Right *Map of the Amazon monsoons, which cover over 3 million square miles in total.*

SOUTH AMERICA, AMAZON RAINFALL

KEY

	1 inch (25 mm)
	2 inch (50 mm)
	3 inch (75 mm)
	4 inch (100 mm)
	6 inch (150 mm)
	10 inch (200 mm)
	12 inch (300 mm)
	16 inch (400 mm)
	20 inch (500 mm)
	40 inch (1000 mm)

ASIAN MONSOONS

Three-quarters of Bangladesh is a deltaic region formed by the sediments brought in by the Ganges, Brahmaputra, and Meghna rivers, which are all fed by the summer monsoons. More than half the country lies less than 13 feet (5 m) above sea level, leaving it prone to frequent flooding. During a normal summer monsoon a quarter of the country is flooded. The monsoon floods worsened throughout the 1990s, and in 1998 three-quarters of the country was flooded for two months, causing billions of dollars worth of damage and thousands of deaths.

But along with destruction, these floods bring life, as the water and silt irrigate and fertilize the land. The fertile Bengal Delta supports one of the world's most dense populations—over 110 million people in 55 thousand square miles (140,000 square km).

Top *India and Bangladesh receive from 16 to over 32 inches (400 – 800 mm) of rainfall between June and August each year. Much of this staggering amount of rain ends up in Bangladesh as three of the largest rivers in the Indian subcontinent flow into the Indian Ocean from here. The monsoon in South-East Asia is also very vulnerable to El-Niño events. As this map shows the percentage increase and decrease in monsoon rainfall during El Niño can be extreme, affecting everything from the occurrence of floods to a lack of water for crops.*

Right *For two months every year the monsoons come to India and everyone has to adapt to near-constant rainfall.*

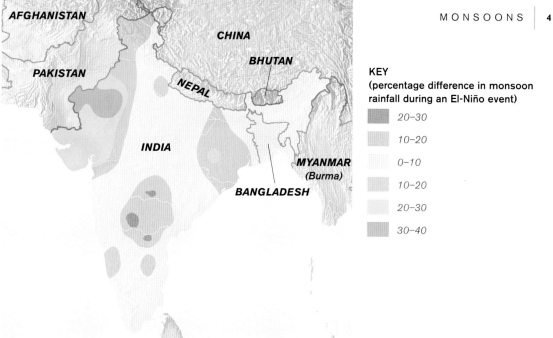

KEY
(percentage difference in monsoon rainfall during an El-Niño event)

- 20–30
- 10–20
- 0–10
- 10–20
- 20–30
- 30–40

EL NIÑO-SOUTHERN OSCILLATION (ENSO)

Changing global weather patterns

Changes in the direction and intensity of currents and winds in the Pacific Ocean can have a dramatic effect of the climate of other parts of the world–but just how well do we understand why it happens?

Far right *Common effects on the weather of El Niño and La Niña conditions around the world.*

Above *The El Niño event of 1998 was so strong that it affected areas as far away as California. We can see from this picture how much greener the San Francisco Bay area is as a result of the increased rainfall.*

This phenomena was originally known as El Niño (Spanish for "Christ Child," since it usually appears at Christmas). It is now called the El Niño-Southern Oscillation (ENSO) because we have come to understand that the climate in the Pacific Ocean, where it occurs, alternates between two opposite extremes, El Niño and La Niña. ENSO typically occurs every three to seven years.

An ENSO event can last for anything from several months to over a year, and a prolonged one can cause severe climate changes all over Earth. Droughts in East Africa, Northern India, North-East Brazil, Australia, Indonesia, and the

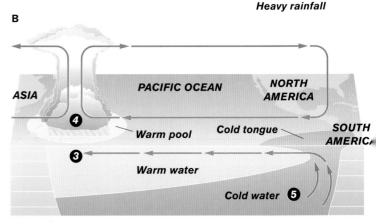

A

ASIA   PACIFIC OCEAN   ❷   NORTH AMERICA

SOUTH AMERIC

Warm pool

Warm water   ❶

Cold water

Heavy rainfall

B

ASIA   PACIFIC OCEAN   NORTH AMERICA

❹   Warm pool   Cold tongue   SOUTH AMERIC

❸   Warm water

Cold water   ❺

Southern United States, together with heavy rains in California, Eastern Central Africa, Sri Lanka, and parts of South America, are just some of its effects.

ENSO has been linked to dramatic changes in weather patterns throughout the world, including the position and occurrence of hurricanes in the Atlantic. It is thought that the poor prediction of where Hurricane Mitch made landfall was the result of ENSO conditions not being taken into account.

Left *Schematic Diagram of the difference between El Niño and La Niña conditions*

A *EL NIÑO CONDITIONS (1) Warm surface water in the Western Pacific Ocean moves across to the center of the Pacific Ocean. (2) The warm rising air which it causes is, thus, much closer to South America. (3) Consequently the Trade Winds are much weaker and the upwelling of cold nutrient-rich water almost ceases. This changes the direction of the jet streams which upsets the weather in North America and Europe.*
B *LA NIÑA CONDITIONS (4) Warm surface water in the Western Pacific causes warm, moist air to rise from the ocean surface spawning thunderstorms. (5) This rising air sucks in the Trade Winds, pulling them across the whole of the Pacific ocean. As the Trade Winds move away from South America they push away the surface water of the ocean forcing cold nutrient-rich water up from much deeper.*

EL NIÑO EFFECTS

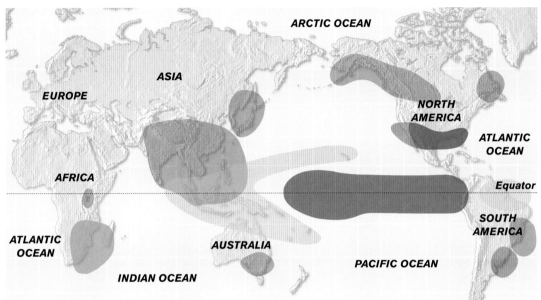

LA NIÑA EFFECTS

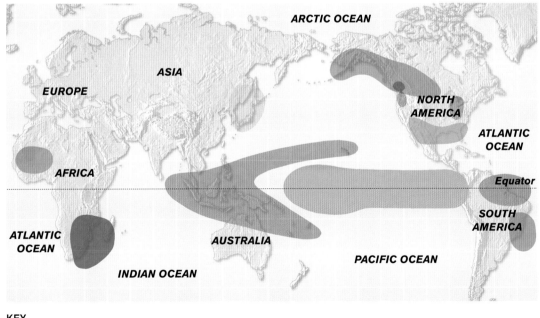

KEY

Warm	Cool	Wet	Dry
Wet and cool	Wet and cool	Dry and warm	Dry and cool

Having already experienced three El Niño events in the 1990s (1991–1992, 1993–1994, and 1994–1995), climate researchers were surprised by the beginning of another El Niño event in 1997. It turned out that this was "the big one," even bigger than the El Niño event in 1982–1983. For all the advance warning given, the 1997 El Niño devastated much of the world with flood and fire, and was blamed for 23,000 deaths and for causing 32 billion US dollars worth of damage worldwide.

EL NIÑO

El Niño occurs when the warm surface water in the Western Pacific Ocean moves eastward to the center of the Pacific Ocean—bringing the warm column of rising air much closer to South America. As a result, the Trade Winds and the ocean currents crossing the Pacific Ocean are weakened. The massive shift in ocean currents and the position of the rising warm air changes the direction of the jet streams, upsetting the weather in North America, Europe, and the rest of the world.

Along the western coast of South America, the weakening of winds and currents has dramatic consequences. Usually, strong winds push the ocean currents away from the coast and, to fill the space left, water is pulled up from beneath. This water is rich in nutrients which provide food for sea life and sustain the large fish catches on which many communities rely. During an El Niño period, weaker winds mean that ocean currents are not pulled as far from the coast. Less water and nutrients are upwelled, resulting in a severe reduction in fish catches and putting economic stress on the region.

LA NIÑA

La Niña is an exaggerated version of the "normal" conditions. Normally, there is a warm pool of water in the Western Pacific Ocean, which is kept in place by strong westerly winds and ocean currents. During a La Niña period, the temperature difference between the Western and Eastern Pacific Ocean becomes extreme and the westerly winds and ocean currents are enhanced.

La Niña's impact on the world's weather is less predictable than El Niño's. During an El Niño period

Above *Australia often undergoes severe drought during El Niño conditions, as the rains have shifted to the center of the Pacific Ocean.*

the Pacific jet stream and storm tracks grow stronger and straighter, making it easier to predict the effects. Conversely, La Niña weakens the jet stream and storm tracks, making them more irregular. The behavior of the atmosphere and, in particular, of storms, is more difficult to predict.

In general, where El Niño is warm, La Niña is cool; and where El Niño is wet, La Niña is dry. While El Niño conditions and their seasonal impacts look very different from normal, La Niña conditions often bring "typical" winters—only more so. Stronger-than-normal winds pull the surface ocean currents further away from the coastline, allowing even more nutrient-rich water to be upwelled. This results in increased fish catches for the season.

WHAT CAUSES EL NIÑO?

This is a chicken-or-egg situation. Which comes first? Does the westward ocean current across the Pacific Ocean reduce in strength, thereby allowing the warm pool to spread eastward moving the wind system with it? Or does the wind system relax in strength, reducing the ocean currents and in turn allowing the warm pool to move eastward?

Many scientists believe that there are very large waves in the Pacific Ocean which move between South America and Australia/South-East Asia. These waves have a very low height but are extremely long. One way to visualize them is to think of the waves you create when you get out of the bath, with the ends of the bath representing South America and Australia. Over time these waves, bouncing slowly across the Pacific Ocean, help shift the ocean currents producing either an El Niño or La Niña period.

HOW BAD IS EL NIÑO?

It depends where you live. It's terrible news for Ecuador and Peru as these countries suffer terrible flooding that decimates the fishing industry. It is bad news in Southern California, where it brings heavy rains and causes extensive flooding. It also brings very dry conditions to Southern Africa, Indonesia, and Northern Australia, causing droughts and extensive forest fires. If you happen to be in North America or Europe, however, you can thank El Niño for milder winters.

Left *During El Niño conditions, Australia is even drier than usual and forest fires are common.*

Right *California usually experiences unusually heavy rains during El Niño—so much so that the land becomes saturated and falls, bringing down hillside houses with it.*

Below right *During the 1998 El Niño, conditions became so dry that Lake Poopó in the Bolivian high Andes, which was full in March 1997, was almost completely dry by May 1998.*

March 1997

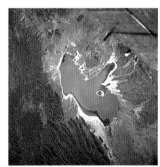

August 1997

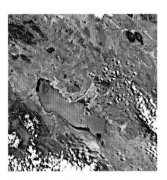

November 1997

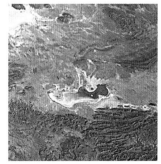

May 1998

BUILDINGS
High-rise horrors

Although many buildings are now designed to withstand earthquakes, surprisingly little work has been done to mitigate the effects of hurricanes and tornadoes—and the consequences can be devastating.

Below left As this diagram shows, when wind encounters a high building it has three options: to go around the building, over it, or through. High-rise buildings are constructed with this in mind.

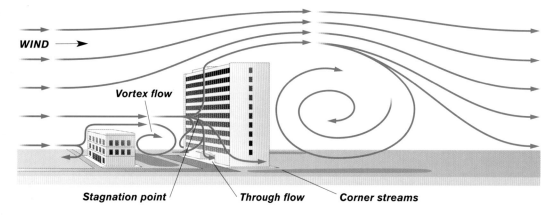

WIND →

Vortex flow

Stagnation point · Through flow · Corner streams

Storms can damage buildings in three main ways. First, winds traveling at high speed can crash directly into buildings causing severe structural damage. Second, flying debris picked up by the storm can be thrown against buildings. In severe hurricanes and tornadoes with high wind speeds, cars, trees, and even parts of other constructions can be picked up and hurled through the air. The third danger is air pressure, which, particularly in the eye of a hurricane or tornado, can be very low. The pressure difference between the storm and the building's interior can cause buildings to explode outward. In the case of tornadoes, the suction from the twister can pull buildings away from their foundations and fling them literally hundreds of feet into the air.

HURRICANE ALICIA
On August 17–18, 1983, Hurricane Alicia struck the business district of Houston, Texas. Luckily, weather predictions provided enough warning for the area to be evacuated, with no personal injuries. The buildings were also hurricane strengthened, so although they swayed under the winds, none of the frames were damaged.

But the storm did destroy the buildings' glass cladding and windows, despite the use of double panes, heat strengthening, and tempered, thicker plates. It seems that the wind stress found minute flaws in the glass and exerted huge pressure, causing some windows and glass cladding to fail. Once a few windows had failed, the winds then battered these fragments against other windows. The 71-storey Allied Bank building had nearly 400 broken windows and 1,200 broken lights, which ended up as over 3,000 sharp glass fragments tossed about in the storm like deadly knives.

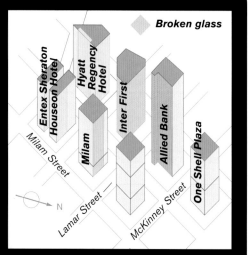

Broken glass

Entex Sheraton Houseon Hotel · Hyatt Regency Hotel · Inter First · Allied Bank · Milam · One Shell Plaza · Milam Street · Lamar Street · McKinney Street

N

A schematic diagram of the windows that were completely blown out by Hurricane Alicia when it hit Houston, Texas, on 17–18 August 1983.

THE "STREET-CANYON" EFFECT
High-rise buildings are prone to storm damage because of their size and height. This can not only prevent people from being evacuated quickly and safely, but also creates the danger of a street-

canyon effect. This occurs when wind is channeled through a cluster of skyscrapers standing close together, causing the speed of the winds to increase. Although we are well aware of the dangers, when models of skyscrapers are tested in wind tunnels, they are generally tested in isolation, so the street-canyon effect is rarely taken into consideration.

BUILDING EVACUATION

Although computer models are used to plan evacuation procedures, they usually assume that people are familiar with evacuation procedure and will act in a predictable manner. However, there is a great difference between a fire drill and a real emergency, especially when a hurricane or tornado hits a high-rise building without warning. Most people's instinct is to leave a building during a natural disaster or emergency, but flying glass and the street-canyon effect in the modern city at times makes this a deadly proposition. The priority for someone trapped in a storm in a high-rise building is to get to the center of the building, away from the windows. Stairwells are usually the safest place.

Computer model of the evacuation procedures from a skyscraper during a fierce storm. The first priority is to get people to move away from the windows as these will explode. If possible, find a room without windows and shelter there. Otherwise move to the nearest stairwell (never use an elevator) and leave the building as quickly as possible. Notice how the main corridors and stairs become bottlenecks. Under these conditions it is of utmost importance that people stay calm—panic leads to chaos.

53rd floor

EVACUATION ROUTE KEY

person

injured person

struck person

struck and injured person

Right *These buildings were unfortunate enough to lie in the path of a hurricane. Imagine witnessing that firsthand!*

LIGHTNING
The bolt from the sky

Lightning is the thunderstorm's most dangerous threat, resulting in some 1,000 deaths every year around the world, and a further 5,000 injuries, many serious.

Although lightning strikes are a worldwide phenomenon, only in the United States has an attempt been made to quantify the hazard they pose to human life. Studies show that while, on average, individual lightning strikes kill more people than other obvious storm catastrophes such as tornadoes and hurricanes, they cause fewer injuries. The table below shows the number of deaths and injuries per ten million people in a total U.S. population of 240 million.

	Lightning	Tornado	Hurricane
Death	5.43	5.24	2.52
Injuries	10.95	90.48	119.52

WHAT CAUSES LIGHTNING?

A lightning strike (or stroke) is the result of a violent atmospheric discharge of electricity. Turbulence caused by rapidly moving air within a thundercloud leads to a build-up of static charge. Positive charges concentrate in the upper part of the cloud and negative charges at the bottom. The negative charge nearest the earth induces a mainly positive charge in the ground below it. As the charge difference between the top and bottom of the cloud becomes large enough to overcome air resistance, electrons discharge toward the earth in a forked path, traveling at about 60 miles (96 km) per second and heating the surrounding air to temperatures of up to 50,000°F (28,000°C).

People and animals caught in the path of a strike risk electrocution. Other materials are simply vaporized. The phenomenal temperatures start fires, and the associated burst of electromagnetic energy can damage electrical and electronic equipment for huge distances around.

HOW LIGHTNING ORIGINATES

A (*1*) *Positive charges gather at the top of the cloud.* (*2*) *Negative charges gather at the bottom.* (*3*) *This induces a positive charge in the ground below.* (*4*) *Negative electrons flow toward the ground in a stepped leader.* (*5*) *Positive electrons stream up to meet it.*
B (*6*) *The stepped leader and positive electrons meet, creating a conductive path or channel from cloud to earth.*
C (*7*) *Electrons rush to the ground creating the visible stroke, and the return stroke reaches back into the cloud, relieving it of its charge.* (*8*) *The superheated air around the stroke expands rapidly, causing the shockwaves that we hear as thunder.*

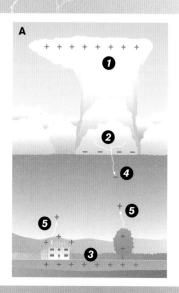

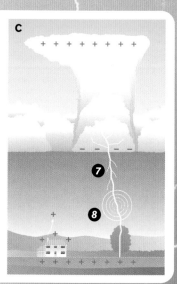

EMBRACE OF DEATH

In 1999 two Japanese tourists were killed in London's Hyde Park by a freak lightning strike as they sheltered from the storm under a tree. The electricity from the lightning bolt was conducted by the metal wires in their bras and stopped their hearts.

LIGHTNING FACTS

Most people struck by lightning are not killed, and two out of three make a full recovery. There is no truth in the theory that people struck by lightning continue to carry an electrical charge. It is perfectly safe to touch them, and essential to give them immediate first aid.

IN THE EVENT OF LIGHTNING:

- Take cover in a building if possible.
- Stay clear of electrical conductors, such as telephone wires, plumbing pipes, and other metal objects.
- Do not shelter under a tree.
- Avoid tall structures, especially metal ones, and water.

REMEMBER:

- The lightning threat does not pass when rain ends. Lightning often strikes when it is not raining or just after rain has stopped.
- Cars, trains, and airplanes are safe environments because, when struck, the metal body conducts the electricity around and away from the occupants.

Top left *The fact that the electrical charge seeks the shortest path to earth explains why trees and other tall structures are at particular risk.*

Left *Multiple strokes are common in a violent storm, such as this one over Seattle, USA.*

WINTER STORM DAMAGE

The devastating effects

The consequences of any snow-, hail-, or ice storm can be dire, but it is only when a roaring avalanche comes crashing down a mountain that the most serious effect of these winter storms is witnessed.

Right *Avalanche coming down the side of Mount Rakaposhi in the Northern Territory of Pakistan.*

Far right *Avalanches can flatten or bury houses in seconds.*

ICE STORM OF THE CENTURY

In early January 1998, as warm, moist air from the Gulf of Mexico rode over a layer of Arctic air, rain fell for five days. Falling through the cold air, it froze instantaneously on every exposed surface. Many of the northern states of the United States declared an official emergency, but it was Canada that took the brunt of the storm. From eastern Ontario to New Brunswick and Nova Scotia, the weight of ice crushed trees and toppled steel electrical power transmission lines. At the height of the disaster eight million people were left without electricity, and many were still without power a month later. The storm killed 35 people and property damage ran into billions of US dollars. In Quebec and Ontario, where 90 percent of the world's maple syrup is produced, millions of trees were destroyed. It is predicted that it will take another 30 years for maple syrup production to return to normal.

Above *Power cables brought down in Canada by the weight of ice*

A blizzard may pose a direct threat to your safety because of strong winds and poor visibility, but it is generally the kind of weather you can shelter from. Hail and snow rarely cause actual physical injury. They can, however, have a severe economic impact: in the United States, hail and snow storms cause the loss of two percent of the country's crop and 760 million US dollars damage every year.

No winter storm, however, causes as much damage as an ice storm. During long periods of freezing rain, the ice that accumulates crushes everything under its weight. Electrical power systems that depend on overhead lines are no match for a serious ice storm. Power lines droop under the weight, causing the poles to snap. With power down and transportation almost impossible, the effects of an ice storm linger for days.

AVALANCHES

Capable of crushing everything in its path, the avalanche is the most devastating effect of snowstorms. When snow accumulates on slopes of between 25–40° (any shallower and the gradient is not steep enough to make the snow fall; any steeper and snow can't accumulate), it becomes compacted and can start to slide without warning. When this happens the avalanche begins, producing a wall of up to a million tons of snow that can hurtle down the slope at speeds approaching an incredible 200 miles (320 km) an hour.

AVALANCHE PREVENTION

Snow bridges

Snow fences

Avalanche shed

Splitting wedges

Deflecting walls

Avalanche breakers

The fastest avalanche measured to date struck at Glarnisch, Switzerland on March 6, 1998. It reached 217 miles (350 km) per hour and traveled 4.3 miles (6.9 km) in just over a minute.

Avalanches are very common. There are about 100,000 each year in the American Rockies alone, mostly in uninhabited areas where there is no threat to human life. It is when avalanches occur close to towns and roads that they do the most damage. With the ever-increasing popularity of skiing and snow boarding, the population in areas threatened by avalanches is growing. In the small village of

Galtur, Austria, avalanches killed 38 people during the severe snow storms that hit the Alps in 1999.

The effect of avalanches can usually be minimized by a combination of barriers and the controlled use of explosives, which

set off small avalanches, thus preventing them growing to a dangerous size. This strategy is employed at all ski resorts and high-risk areas. The figure above shows the various structural methods of controlling avalanches.

Above *Structural methods of preventing avalanches from causing death and destruction.*

FLOOD

The waters of life—and death

Causing billions of dollars worth of damage and high numbers of fatalities, floods have a far-reaching impact. Frighteningly, they are becoming more widespread and more frequent.

In terms of yearly loss of life, just under 40 percent of fatalities from natural hazards are caused by floods. It has been estimated that between 1964 and 1982 floods claimed 80,000 lives and affected 221 million people. These deaths are not evenly distributed around the world: from 1986 to 1995 Asia suffered 44 percent of the world's flood disasters and 93 percent of the fatalities. China is the most frequently flooded country, with repeated disasters throughout its history. One flood, recorded in 2297 B.C., was so large that the Yellow, Wei, and Yangtze rivers burst their banks and turned the Northern Chinese plains into an enormous inland sea.

FLASH FLOODS

Flash floods occur as a result of motionless or slow-moving thunderstorms, which produce nearly constant rain. The amount of water falling, usually over 2 inches (5 cm) per hour, means that the ground very quickly becomes full of water and is unable to absorb any more. This water then runs along the surface straight into rivers, causing their banks to break and flood. Flash floods can also be caused when there has been a lot of snow build-up.

Flash floods are so destructive not only because they have so much energy and contain a great deal of mud, sediment, and boulders, but also because there is little warning. The most troubling thing about flash floods is that they can affect areas many hundreds of miles from the source of the water.

Right *In 2000, the worst floods for a hundred years hit Mozambique, washing away both mud and stone buildings and livestock.*

Below right *Bangladesh experiences monsoon floods every year. Sometimes, as in 1998, the floods are so bad that the whole country grinds to a halt.*

Below *The Somme area of northern France, which experienced severe floods in 2000.*

MILLENNIUM MAYHEM

The turn of the century has seen some of the worst flooding in decades. Many scientists suggest this could be due to the increasing effects of global warming (see pp. 104–105).

- In 1998, three-quarters of Bangladesh was under water for two months.
- In 1998, large parts of China flooded, killing 3,700 people and causing $30 billion worth of damage.
- In November 2000, the north of England experienced two once-in-thirty-year floods. York was particularly badly affected: the River Ouse rose over 16 feet (5 m), exceeding the previous record set in 1625.
- In winter 2000–2001, huge floods in Italy destroyed historic villages.
- In 2000 and 2001, Mozambique experienced its worst floods, which rendered thousands of people homeless.

Above *Britain was hit by the worst floods ever recorded in 2000, with rivers rising by as much as 16 feet (5 metres).*

WHY DO PEOPLE LIVE ON FLOOD PLAINS?

Floods are also a positive part of human life, carrying sediment that revitalizes the fertility of the soil for crops and agriculture. This is why many flood plains have a long history of human settlement. The Euphrates, Ganges, Indus, Nile, Tigris, and Yangtze rivers have all supported major civilizations because of regular floods. There are also many activities, besides agriculture and fishing, that depend on rivers, such as log transport, barge traffic, and power stations, which require a source of cooling waters. Many flood plains and their rivers are important corridors for road, rail, and water transportation. Moreover, because they are flat, flood plains make the perfect site for the construction of buildings.

STORM SURGE AND COASTAL EROSION

At the mercy of the elements

Coastlines all around the world are under threat from severe storm conditions. Global warming is a major contributing factor.

STORM SURGE

The danger of storm surge is greatest when a hurricane makes landfall during high tide; this can produce wind-whipped waves as high as 15 feet (4.5 m)–a potentially lethal combination for harbors and seaside communities.

Some storm surges are even greater than this. Hurricane Camille, a Category 5 hurricane that struck Mississippi in 1969, produced a 25-foot (6-m) surge. In 1973 a cyclone hit Bangladesh producing a 23-foot (5.5-m) surge that drowned 300,000 people.

NO SAFE HAVENS

Storm surges are produced by many different types of storms and can affect areas that are not under threat of hurricanes and cyclones. On February 1, 1953, a North Sea storm surge hit the Eastern United Kingdom and Holland, damaging 400,000 buildings and killing 1,835 people and 47,000 animals.

And the height of the wave is not the only problem. When a hurricane hits, the large dome of water that crashes over the shore can be 50–100 miles (80–160 km) wide. Three things control a storm surge: wind speed, water depth, and the very low pressure within a storm which causes the sea to rise. The stronger the winds, the shallower the offshore water; the lower the pressure in the storm, the higher the storm surge.

COASTAL EROSION

Although very few fatalities have been recorded as a result of coastal erosion, the speed at which it occurs can cause serious economic problems—particularly when farmland or heavily populated areas are threatened. With scientists predicting rising sea levels and more severe storms as a result of global warming, coastal erosion looks set to become an increasing problem.

Storms contribute to coastal erosion in three main ways:

- Heavy rains soak into the ground, making cliffs unstable and prone to collapse.
- Increased wind speeds during storms increase the height of the waves pounding the coastline.
- Increased wave height can lead to a storm surge which floods the land behind the coast. As the water is sucked back into the sea, the coast is eroded.

BEACHY HEAD

At 570 feet (171 m) high, the gleaming, white chalk cliffs at Beachy Head, in Southern England, provide a fine viewpoint over the English Channel and have been a landmark for centuries.

On January 11 1999, 100,000 tons of rock collapsed and bridged the space shown above between the cliff and the lighthouse. This was the result of heavy rains and huge waves pounding the soft chalk of the cliff.

STORM SURGE

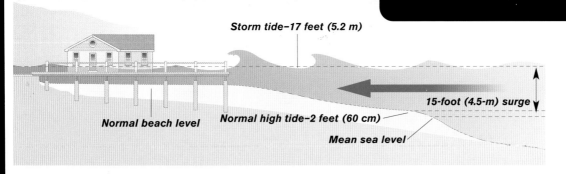

Storm tide–17 feet (5.2 m)

Normal beach level

Normal high tide–2 feet (60 cm)

Mean sea level

15-foot (4.5-m) surge

Above left *Storm surges are one of the main causes of catastrophic cliff failure.*

Opposite page *When waves crash against the shore, they are pushed upward and further inland than normal.*

HURRICANES TO REMEMBER

Gilbert and Galveston

Two deadly hurricanes almost a century apart—with widespread devastation in both instances. The difference? Improved forecasting in 1988 meant that the death toll stopped at 45 for Hurricane Gilbert. For Galveston Hurricane, 88 years earlier, it had topped 8,000.

"Jamaica has been highly active in promoting evacuation and sheltering plans, and here loss of life nationally was only 45, including 11 people shot by the police as presumed looters."

"National Disasters," David Alexander (1993) on the subject of Hurricane Gilbert in 1988

THE STORM

Hurricane Gilbert hit Jamaica on September 12, 1988, and made landfall at a speed of 15 miles (25 km) per hour. The eye of the storm was over 24 miles (40 km) in diameter and wind speeds of 93 miles (150 km) per hour were sustained while gusts hit at 139 miles (225 km) per hour. It was one of the worst storms recorded in the Caribbean.

THE DAMAGE

Forty-five people died in Hurricane Gilbert. The loss of life might have been much worse, but because Jamaica is frequently in the path of hurricanes, it has been very active in promoting evacuation and shelter plans.

Nevertheless, Hurricane Gilbert is a prime example of how simple damage can have complex effects on a society. One in four houses were damaged, mostly due to roofs being ripped off, allowing

HURRICANE GILBERT: CARIBBEAN, MEXICO, AND U.S.A.

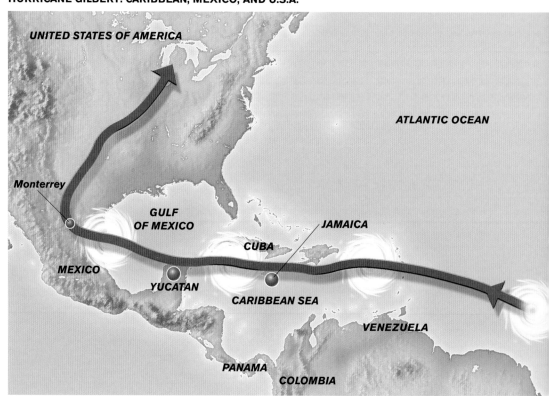

Above *Pathway of Hurricane Gilbert in 1988*

GALVESTON HURRICANE, 1900

Before the age of satellites and hurricane hunter aircraft, tracking hurricanes was very much a hit-and-miss affair. On September 8, 1900, the people of Galveston, Texas, were caught unawares by one of the worst hurricanes of the 20th century. A 20-foot (6-m) high storm surge completely swamped Galveston, and all through the night hurricane winds smashed the rafts that the few survivors clung to. The next morning, Isaac Cline of the Galveston weather service office saw "one of the most horrible sights that ever a civilized people looked upon." Because whole families were simply washed away the exact death toll will never be known, but it is estimated that more than 8,000 people died, making the Galveston Hurricane one of the deadliest natural disasters in U.S. history.

wind and rain to ruin interiors. In one case a roof was ripped off a church in which 400 people were sheltering. Jamaican buildings were not all built to the same standard, so the damage hit some structures harder than others. Over 500 schools were badly damaged, ten hospitals were affected, and one insurance company even had its records destroyed by rain, causing chaos for claimants who wanted to rebuild after the hurricane.

Communications, too, were badly affected. When the roof of Jamaica's main telephone exchange was blown off, rain destroyed the switchboard, cutting all telephone lines. Radio and television masts were brought down by the terrible winds, and this prevented authorities from organizing an efficient help and rescue operation. There was also widespread damage to crops and industry, with economic loss to the island totaling 3 billion US dollars.

Top left *Satellite image of Hurricane Gilbert approaching Jamaica on September 12, 1988.*

Above *Although Hurricane Gilbert had already lost a lot of its strength by the time it reached Mexico, it was still capable of causing significant damage.*

Right *Hurricane Gilbert's winds were so strong that it threw this airplane into a tree in Kingston, Jamaica.*

THE LESSONS

Prior to Hurricane Gilbert, Jamaica had made great advances in planning how to save lives during the worst hurricanes. But the aluminum sheeting often used for roofs in Jamaica is easily torn away from its fasteners during a hurricane. As a result, the law has been changed and the Jamaican building code now states that a structure must be able to resist a three-second gust of 125 miles (200 km) per hour.

HURRICANES TO REMEMBER

Andrew

Hurricane Andrew left 53 people dead—a toll that, thanks to accurate forecasts and early warnings, was much lower than it might have been. Financially, however, it was the costliest natural disaster in U.S. history.

"Hurricane Andrew in 1992 was the most devastating hurricane to strike the United States."

John Cox (2000)

THE STORM

In terms of the damage it caused, Hurricane Andrew, which came ashore south of Miami Beach, Florida, on the morning of August 24, 1992, is the most devastating hurricane ever to have hit the United States. It crossed Southern Florida in three hours, completely leveling the town of Homestead, before continuing westward into the Gulf of Mexico and then turning northward into Louisiana two days later.

Wind gusts of 175 miles (282 km) per hour were recorded, and there was a sustained wind speed of 145 miles (233 km) per hour, making Andrew a Category 4 hurricane. In Biscayne Bay, the storm surge was a frightening 17 feet (5 m) high.

Fortunately, due to accurate forecasting, the number of fatalities was low. The financial damage, however, was a different story: as so many people lived in its path, the property damage bill was an immense 30 billion US dollars. If the hurricane had hit 20 miles (32 km) further north in the densely populated area of Miami, the damage bill might have doubled.

Below left *Hurricane Andrew tore through this trailer park in Homesteads, Florida, flattening nearly 90% of the trailers.*

Below *Miami after Hurricane Andrew had passed.*

THE AFTERMATH

Unlike most hurricanes, Andrew's winds caused more damage than its storm surge. More than 200,000 homes and businesses were damaged or destroyed, leaving more than 160,000 people homeless. One million people in Florida and 1.7 million in Louisiana and Mississippi were evacuated from their homes thanks to early warnings and excellent evacuation procedures. For all its power, Andrew was a relatively small hurricane, the rain band reaching out only 100 miles (160 km) from the eye.

Above *Satellite picture of Hurricane Andrew spinning across Florida and the Gulf of Mexico.*

Right *A view of Cutler Ridge, Florida, after Hurricane Andrew had passed.*

HURRICANES TO REMEMBER

Mitch

"I have seen earthquakes, droughts, two wars, cyclones, and tidal waves. But this is undoubtedly the worst thing I have ever seen."

Cardinal Miguel Obando y Bravo of Managua, Nicaragua, in response to the terrible devastation caused by Hurricane Mitch

THE STORM

On October 28, 1998, virtually without warning, the most destructive hurricane in 200 years hit the poorest countries in Central America. Honduras, Nicaragua, El Salvador, and Guatemala were battered by 180-mile (290-km) per hour winds, and more than 23

HURRICANE MITCH, CENTRAL AMERICA

Left *Predictions of the pathway of Hurricane Mitch were extremely inaccurate, as the strength of the El Niño-Southern Oscillation in the Pacific Ocean had not been taken into account. Hence Hurricane Mitch veered westward into Central America instead of continuing northward as predicted.*

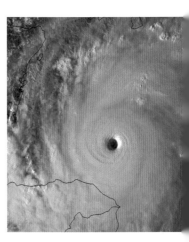

21:45, 26 October 1988

inches (60 cm) of rain every day. Hurricane Mitch packed two killer punches: widespread destruction because of the high winds, and massive floods in the wake of such torrential rain. We will never know how many people died as a result of this devastation, as many bodies were buried under thousands of tons of mud. Estimates put the death toll at over 20,000 people, with up to two million men, women, and children left homeless. Could the effects have been predicted?

THE PREDICTION

The National Hurricane Center and the Hurricane Research Center, both based in Miami, Florida, admit they got the prediction of Hurricane Mitch wrong. Their computer models told them the storm would turn north-west when it was in the Caribbean Sea. Instead it turned south-west and hammered into the poorest countries on the eastern seaboard of Central America. Scientists now believe the problem with predicting the path of Hurricane Mitch was that they had not taken account of the strength of the El Niño-Southern Oscillation in the Pacific Ocean (see pages 42–43), which may have helped drag Hurricane Mitch across the land strip into the Pacific Ocean.

THE WARNING

Scientists were thwarted in their attempts to alert the relevant government authorities of the Central American countries by poor communications infrastructure. When the warning did get through, it predicted only the wind speeds and made no mention of the expected rainfall, which is what caused most of the damage.

THE AFTERMATH

The worst hit was Honduras, a small country of only six million inhabitants. The Hamuya River, normally a calm stretch of water

Above *Hurricane Mitch, classified as a Category 5 hurricane on the Saffir-Simpson Scale. It killed 20,000 people, made over 2 million people homeless, and caused over 2 billion US dollars worth of damage in Central America.*

Below *This sequence of satellite pictures shows how Hurricane Mitch developed and intensified in strength. The hurricane sustained wind speeds of over 180 miles (290 km) per hour, and more than 23 inches (60 cm) of rain fell every day while it passed over Central America.*

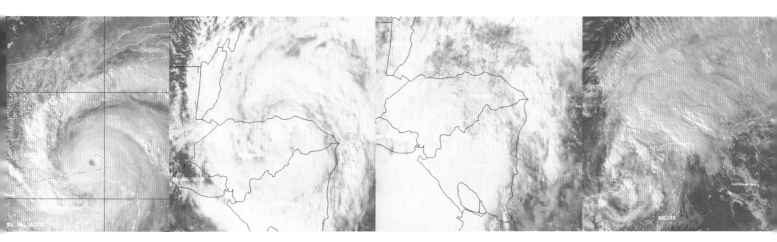

20:21, 27 October 1988　　*20:15, 29 October 1988*　　*20:45, 30 October 1988*　　*20:44, 3 November 1988*

about 200 feet (60 m) wide, rose by 30 feet (9 m) and became a raging torrent 1,500-feet (450-m) wide that ripped up trees as tall as a seven-floor apartment block. Eighty-five percent of the country ended up under water. Over 100 bridges, 80 percent of the roads, and 75 percent of its agriculture were destroyed, including most of the banana plantations. The total repair bill for the damage done by Hurricane Mitch is estimated at over 2 billion US dollars.

THE LESSONS

The inadequate warning of the hurricane's destructive capabilities contributed to the devastation in Central America. But the most important lesson is that extremely poor countries, such as those hit by Hurricane Mitch, just don't have the money to cope with the damage inflicted by natural disasters. In Central America this is exacerbated by crippling international debt. In Nicaragua, for example, the government struggled to find money to fuel its fleet of aging Soviet-built helicopters—essential in delivering aid and evacuating the homeless.

The international community was also slow to respond to the crisis: offers of equipment, money, and manpower only started to come in a week after the storm had hit. Eventually the developed world came up with a rescue package worth 100 million US dollars—a drop in the ocean compared to the 2 billion dollars required to rebuild the area.

"I swam and swam, trying to save my son, trying to get somewhere dry. And then I realized I was already in the sea. I was begging God to let someone find me and rescue me. But there was no one. No one saw me. The worst part for me was after being with my whole family, with my children, my husband, that I could be so alone in the sea without seeing anyone."

Laura Arriola de Guity, who was swept out to sea by the floods caused by Hurricane Mitch. She was found six days later, 75 miles (120 km) from her home. Her entire family perished in the disaster.

Top right The mud slides caused by ten days of intense rainfall from Hurricane Mitch swept through towns and villages such as Choluteca in Southern Honduras, washing away homes and indiscriminately dumping people's possessions over a huge area.

Left The sustained 180-mile (290-km) per hour winds in Hurricane Mitch destroyed many buildings. The country worst hit was Honduras: some of the wind damage in Tegucigalpa is shown here.

Right The second killer blow of Hurricane Mitch was the sustained rainfall of over 23 inches (60 cm) per day. This caused massive mud slides washing away or covering cars, people, and even houses.

TORNADO TALES
The trail of destruction

"I was working that day as a forecaster for the National Weather Service in Louisville, Kentucky. About half an hour after the tornado warning was issued we saw a thunderstorm approaching. As the clouds moved overhead we could see the funnel cloud forming. Suddenly, an instrument shelter that was bolted to the roof deck collapsed in front of our window. The tornado had reached the roof without a visible funnel. This tornado was one of the 148 twisters recorded during the outbreak. For me it was the most spectacular, as it was the first tornado I witnessed."

John Forsing recalls Terrible Tuesday, April 3, 1974

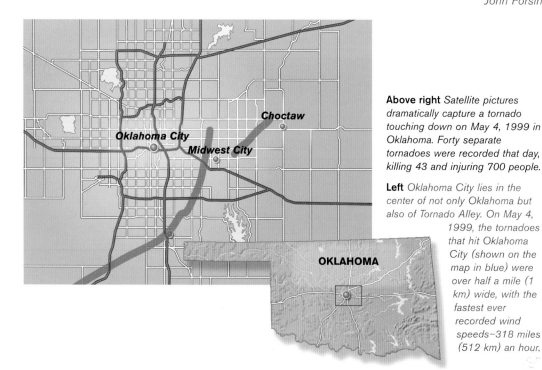

Above right *Satellite pictures dramatically capture a tornado touching down on May 4, 1999 in Oklahoma. Forty separate tornadoes were recorded that day, killing 43 and injuring 700 people.*

Left *Oklahoma City lies in the center of not only Oklahoma but also of Tornado Alley. On May 4, 1999, the tornadoes that hit Oklahoma City (shown on the map in blue) were over half a mile (1 km) wide, with the fastest ever recorded wind speeds–318 miles (512 km) an hour.*

TORNADO TIPS
Do
- Seek shelter in a sturdy structure immediately, preferably underground or on the ground floor
- Stay away from windows
- Lie flat in a ditch or depression if caught in the open
- Protect your head from flying debris

Don't
- Try to outrun a tornado in your car
- Seek shelter under a bridge or overpass—they can act as wind tunnels and make the winds stronger

RECORD BREAKERS

The most destructive tornado ever recorded in the United States was the Tri-State, which raced across Missouri, Illinois, and Indiana on March 18, 1925 at speeds of over 50 miles (80 km) per hour. The consequences were devastating: 695 people were killed and over 2,000 were injured.

The worst multiple-tornado outbreak was on April 3, 1974, known to this day as "Terrible Tuesday." A total of 148 tornadoes touched down over a 24-hour period in 12 states from Michigan to Alabama, killing 309 people and injuring 5,300, and costing 600 million US dollars in damages.

On May 4, 1999 the fiercest tornadoes to hit the United States in a decade left a trail of death and destruction across the states of Oklahoma and Kansas. Nearly 40 separate tornadoes touched down, killing at least 43 people and injuring 700. One tornado that hit Oklahoma City was over half a mile (1 km) wide and had the highest wind speeds ever recorded. Radar readings of the F_5 tornado that hit Moore, Oklahoma, measured wind speeds of an astonishing 318 miles (512 km) per hour. The force of these record winds was so great that paving stones were ripped from the streets. In Oklahoma City alone, 1,800 homes were destroyed, 2,500 were damaged, and there was a repair bill of US$1.2 billion.

"We've got pages and pages of missing people. We've got whole communities that simply aren't there anymore . . . it looks like a huge battle has taken place."

Frank Keating, Oklahoma State Governor, May 1999

FIRST TORNADO EXPERT, AND PANIC ON THE WESTERN PLAINS

The world's first tornado expert was U.S. Army Sergeant John Finlay, who was assigned the job of researching storms in 1882. The U.S. National Weather Service was a new agency under the army's Signal Service, and Finlay's job was to figure out how to predict these killer storms. By 1887, he had a network of 2,400 volunteers reporting tornadoes. He focused his initial ideas for prediction on the dew point and also the winds. However, the Signal Service officials decided that public panic was a bigger threat than tornadoes. Much of the Western Plains were still being settled, and it occurred to the "powers that be" that Finlay's results were not good for real-estate values. So they banned the use of the word tornado in forecasts and abolished Finlay's research unit. The "T" word was officially forbidden until 1938, and it wasn't until the 1950s that the U.S. Weather Service began issuing forecasts for severe thunderstorms and tornadoes.

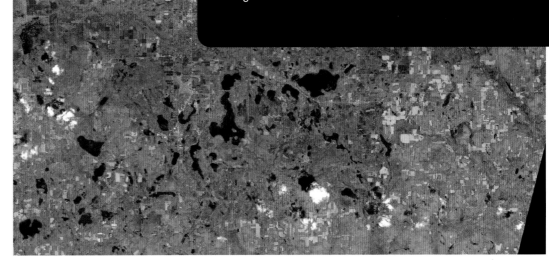

Right *These are the dramatic before (above) and after (below) satellite photographs of a tiny town called Siren, Wisconsin (population 900). On the evening of June 18, 2001 a tornado traveling from west to east killed three people, destroyed homes, and flattened crops. The path of the tornado is illustrated by the paler strip on the lower photograph.*

THE STORM CHASERS

Weather forecasters in Tornado Alley in the United States (see pages 30–31) have a pretty good idea of the conditions that are necessary for severe thunderstorms, and they can often forecast them. Indeed, they can even say whether large hailstones and strong winds are likely.

They can also indicate that a tornado is a possibility during the next few hours or the next day or so–but they cannot forecast it with certainty. The exact conditions within a thunderstorm that are required to create a tornado still remain one of the biggest mysteries in weather science. At the moment the buck stops with the local weather forecast offices. During a severe thunderstorm, the local forecasters know that a lethal tornado could come spinning down out of the dark clouds at any moment, but they cannot be sure until they see it showing up on the Doppler radar screen or a funnel is actually observed.

Hence the National Severe Storms Laboratory has been set up in Norman, Oklahoma, in the center of Tornado Alley. Here, warm air from the Gulf of Mexico meets cold, dry air from the north, creating perfect conditions for the formation of "supercells"– towering, rotating thunderstorms which unleash damaging winds, hailstones bigger than golf balls, and sometimes tornadoes.

Tornadoes are the least understood forms of severe weather, and so scientists are

Left Despite tornadoes being a relatively small type of storm, they are extremely destructive. Let's hope the person in the picture is aware of this! Usually tornadoes form beneath large thunderstorms and reach down to the ground, as can be seen here.

Below A storm chaser stands on his truck. These intrepid scientists use highly sophisticated technology to track the tornadoes.

trying to get near enough to study them. It is from Norman that scientists set out to chase tornadoes every spring.

WHY STUDY TORNADOES?

Of the 1,000 tornadoes reported in the United States each year, only 2 percent are true killers, and twisters spawned from supercells cause about two-thirds of all tornado deaths.

Scientists have already given us a good idea of the general conditions in which tornadoes form. It is the details that still need to be understood–in particular, the mixture of air pressure, temperature, and air circulation in which tornadic thunderstorms evolve. This would help answer a critical question for the millions of people who live in Tornado Alley: Why do some severe thunderstorms produce tornadoes while others do not?

If scientists could find out the conditions in which supercells arise, there would be fewer instances of tornadoes spinning up without warning, and so more people could be evacuated from

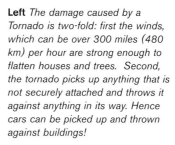

Left *The damage caused by a Tornado is two-fold: first the winds, which can be over 300 miles (480 km) per hour are strong enough to flatten houses and trees. Second, the tornado picks up anything that is not securely attached and throws it against anything in its way. Hence cars can be picked up and thrown against buildings!*

Below left *Though tornadoes only last about 15 minutes, spending just a matter of seconds in any single place, that is enough time to completely destroy a house. In this dwelling in Tulsa, Oklahoma, for example, only the innermost bathroom with no windows was left standing.*

the storm path. Also, there would be fewer of the false alarms which inevitably encourage complacency.

By combining field data from the brave scientists chasing the storms and complex computer models, we are getting closer to understanding how tornadoes are formed. Billions of dollars have already been spent on research, computer models, radars, and satellites, and much progress has been made. Indeed, the lead time for tornado warnings improved from an average of 6 minutes in 1994 to twelve minutes in 1998. This longer warning time has allowed a re-think of the use of public shelters, which were previously impractical as no one

had time to reach them safely. Television meteorologists and other media outlets also play a vital role, continuously broadcasting the location and predicted paths, and generally improving public awareness of tornadoes. And for those of us lucky enough never to witness the destruction first hand, the film *Twister* can give us a good illustration of just what we are missing!

Right *Tornado damage is extremely severe and localized. On May 3, 1999, when an F_5 tornado passed through Moore in Oklahoma, buildings within its 200-yard (200-m) pathway were completely flattened while structures outside this narrow band were left untouched.*

EUROPEAN ALPS
Ski resorts crushed by tragedy

"Left to their own devices, holidaymakers and locals grabbed torches and organized rescue parties. First-aid boxes in cars that could be reached were raided. Unfortunately, emergency equipment had been stored at the fire station, which had lain directly in the path of the avalanche. In the few hours when it really mattered, the lifesaving tools were buried under tons of snow."

Adapted from the Independent on Sunday, February 28 1999

THE STORMS

In February 1999, the European Alps were hit by a massive series of snowstorms and suffered from the heaviest snowfall in living memory. Indeed, the snowfall was so fast and so heavy that many avalanches were set off, leaving a trail of disaster in their wake: more than 70 people were killed, road and rail links were broken, ski resorts were cut off, and tens of thousands of tourists were left stranded throughout the central and western Alps.

Thirty-eight people died when the worst avalanches crashed down on the villages of Galtur and Valzur in the Paznaun Valley, to the southwest of Innsbruck, Austria. The main fall occurred simultaneously on either side of Galtur, and only came to rest after

CENTRAL EUROPE

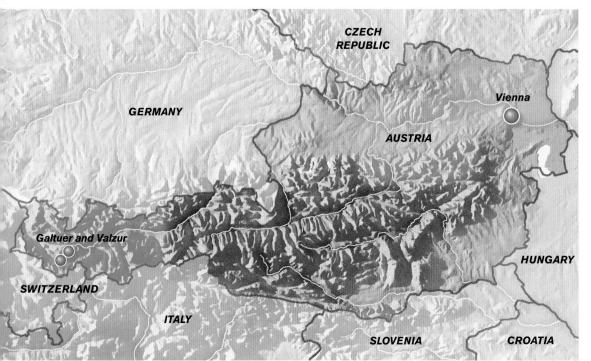

Left *In the center of Europe, its landscape dominated by the Alps, Austria is an extremely popular destination for skiers from all over the world. However, the combination of snow and mountains always brings with it the threat of avalanches.*

it had completely destroyed four buildings.

Unfortunately, the towns had no avalanche barriers because this region had no previous history of huge snowfalls.

The snowdrifts were so large that the only way to get in and out of the Paznaun Valley was by helicopter. The terrible weather delayed the specialist rescue squads from getting to the area until 16 hours after the avalanche. "We could have started the airlift 30 minutes after the avalanche," said Wendelin Weingartner, governor of Tyrol province, "but the best plans amount to nothing in bad weather." When the rescue squad arrived, they found the villagers had spent the night frantically digging with their bare hands to find those buried in the tons of snow.

Further damage was done as the melting snow and heavy rains caused the Rhine River and its tributaries in Germany and Switzerland to flood.

WHY DID IT HAPPEN?

Many scientists suggest that changing snow patterns in the Alps may be the result of climatic changes due to global warming.

Above *Extremely heavy and unexpected snowfall caused many avalanches in the Alps in February 1999. Here, captured on film, is an avalanche roaring toward the town of Zurs in Austria.*

Below *Many avalanche rescues, such as this one in the Alps, can only be carried out by first-aid helicopter. Reaching these isolated areas by other means is a lengthy and dangerous process.*

"We were drinking hot, mulled wine when suddenly it started. The lights went out. It was dark. There was only dust and snow. We got out of there as fast as we could."

Franz Wekno, hotel owner, Galtur, Austria.

FLOODS OF 2000
Disaster strikes

The year 2000 saw the worst flooding for decades in Britain and for 150 years in Mozambique.

BRITISH FLOODS

In October 2000 the worst flooding in decades brought chaos to much of southern England. Millions of dollars' worth of damage was caused as homes and businesses became waterlogged, thousands of people were left stranded, and main roads were blocked. Rivers burst their banks, sending torrents of water cascading through town centers in Sussex, Kent, and Hampshire.

The town of Uckfield, in East Sussex, was one of the worst hit, with 5 inches (13 cm) of rain falling in 12 hours on October 13, leaving the town almost completely submerged. Lifeboat crews were drafted in to rescue stranded residents and workers, and one man was swept away by the fast-flowing floods as he tried to open his shop. He was spotted 20 minutes later, clinging to the banks of the River Uck, and was airlifted to safety by a coastguard helicopter scrambled from Eastbourne, 18 miles (30 km) away.

Severe floods occurred again in November 2000 and February

and December 2001. During these floods over 150 flood alerts were in effect at any one time, demonstrating how widespread the disasters were.

Disruption to transport caused chaos throughout this period, with parts of the train lines from London to the south of England closed for long periods of time in a desperate bid to repair flood damage. At the Eurostar train terminal in Ashford, Kent, some travelers, who had come back from France by train through the Channel tunnel, found their cars had floated away.

Although the floods of October and November 2000 occurred only one month apart, they were so severe that they were classified as floods that only occur once every 30 years.

Right *Although many areas of England were flooded, the area that caught the imagination of the press was the historic city of York. The River Ouse flooded many buildings in November 2000, when it rose by more than 16 feet (5 m).*

THURSDAY, NOVEMBER 1, 2000, ENGLAND

The Archbishop of York paddled round his flooded palace yesterday, wishing that his predecessor had left behind his celebrated "cellar boat." A previous archbishop, the Right Rev. John Habgood, had understood the Bible's obsession with floods, and, with the River Ouse only 10 feet (3 metres) below his study, had invested in a dinghy to help him empty the basement at Bishopthorpe Palace when water poured in, as it does regularly in low-lying York.

The present archbishop, the Right Rev. David Hope, and his staff could not remember a flood like the November 2000 one; indeed, it turned out that the 16-foot (5-m) rise of the River Ouse was a new record, beating the one set in 1625.

The floods also affected people's homes: charity administrator Jo Reilly and her teenage children finally gave up the struggle with the river on Tuesday night, grabbing toothbrushes as the water reached the house's fuse box. "The worst part's yet to come—the cleaning," she said, dangling her rubber boots into the 18-inch- (45-cm-) deep pond in her study. "I know from previous experience that you need to do it at least three times before you can feel it's back to normal." But Jo Reilly and her family can consider themselves lucky that they do not live in the Leeman Street area of York, where the force of the flood sent sewage backing up through drains as storm culverts failed.

But nature has not let Britain off the hook. Since then, in February 2002, gale-force winds of more than 90 mph (140 kph) hit, causing chaos: driving conditions on the motorways were treacherous, and many other roads were affected as trees fell, causing obstructions and traffic diversions. In addition, there were flood alerts on 68 rivers. It seems that floods are going to become a common part of the British winter.

MOZAMBIQUE FLOODS

In February and March 2000, Mozambique experienced its worst floods for more than 150 years. Seven hundred people were killed and 550,000 others displaced. In 2001, four cyclones and record rains brought four successive flood crests and rapidly fluctuating water levels on the Incomati, Limpopo, and Save rivers.

The worst of the tropical cyclones was Eline, which compounded Mozambique's misery after the initial floods when it hit the port of Beira with winds of up to 160 mph (250 kph). The cyclone ripped roofs from homes and flattened traditional mud houses. This was followed eight days later by a second cyclone, Gloria. The cyclones exacerbated Mozambique's problems by producing floodwaters in the neighboring countries of South Africa and Zimbabwe, which spilled into Mozambique's already overflowing rivers.

It was the children who bore the brunt of the flood disaster that devastated Mozambique. The lucky ones were rescued after being forced to abandon their homes. But many others were separated from their parents and an unknown number were stranded and faced the threat of malnutrition.

According to UNICEF's Ian McLeod, in the Mozambican capital, Maputo, of the estimated 900,000 people affected by the floods, 180,000 were children under the age of five. Children account for 46,000 out of the 230,000 displaced people who lost their homes. We will never really know how many children died in this disaster.

PUSHED BACK INTO POVERTY

The floods were an especially cruel blow to Mozambique because, despite being the poorest country in the region, it had become one of the world's fastest-growing economies, at last giving hope to its people, many of whom had been reduced by two decades of civil war to wearing tree bark and eating wild berries. The floods of 2000 and the repeat floods of 2001 pushed back into poverty large numbers of Mozambicans who had recently begun to lift themselves out of it. Just as bad, the flooding dislodged thousands

of landmines from the civil war and carried them into places previously considered safe. Mozambican peasants usually plant some of their crops on low ground, which is damper and more fertile, and some higher up, in case nearby rivers flood. But the floods were so much heavier than usual that not only were the low fields turned to muddy porridge but much of the hillside crop was also destroyed, causing a long-term food crisis in Mozambique.

Left The South African Army and other international relief workers were invaluable in Mozambique's fight against time to rescue people trapped in rural areas.

Below The size of the flood in Mozambique is unimaginable: much of this usually fertile land was covered in water.

CHAPTER 2

Predicting Storms

We stand in awe of the strength and violence of nature. Storms and the floods they bring kill tens of thousands of people every year. However, weather prediction has dramatically improved over the last two decades: we now have the technology and the scientific understanding to predict when and where many of the worst storms will hit, and satellites have revolutionized our ability to forecast the weather accurately. These advances are going to continue as a new generation of super-computers is introduced in the next few years.

Extra warning can save lives, but only if it is combined with improved evacuation procedures, education, and relief and rescue work. Unfortunately, these measures do not come cheap, and much of the world is too poor to be able to invest large amounts in this area.

Above *Through science we are starting to be able to predict the weather more accurately and, more importantly, to predict the occurrence and movement of major storms.*

MEASURING THE WEATHER

Slaves to science

To understand storms, we need to understand the weather and how it changes. The following instruments are used to measure the most important features of weather. These allow us to map changes in temperature, pressure, and cloud cover, all of which help us to predict what will happen over a few days.

Weather station

THE INSTRUMENTS

Thermometer
Thermometers are used to measure temperature (of land, air, lake, and sea). As we have seen, the weather is produced by the need to redistribute heat as evenly as possible over the Earth's surface.

Barometer
Barometers measure atmospheric pressure, which is an excellent prediction tool, as falling pressure often means that a storm is coming.

Lux Meter
Lux meters measure sunlight intensity, which influences the heat of the atmosphere and also what climate we experience. Just think of the difference in temperature when the sun comes out from behind a cloud!

Hygrometers
Hygrometers (or hydrometers) measure relative humidity, or the amount of moisture in the air. Not only does this control the comfort (or discomfort) that we experience at different temperatures, it also indicates changes in clouds, fog, and rain.

Thermometer

Barometer

Anemometer
Anemometers measure wind speed, which provides an indication of the energy in a particular weather pattern. Combined with wind direction, wind speed provides useful information about what the weather is going to do.

Wind Sock
Wind socks (or wind vanes) measure wind direction. This is invaluable when it comes to predicting the weather, as it tells you where your next set of weather is going to come from. For example, in Britain, cold air is associated with winds from the northeast, while rain usually comes from the west.

Rain and Snow Gauges
Both rain and snow are difficult to measure since winds cause precipitation to be unevenly deposited. However, having an indication of rain- and snowfall is important, as it provides a means of assessing both flood and avalanche risks.

Rain gauge

THE INSTRUMENT CARRIERS

Weather Plane

Accurate weather predictions depend on knowing what the atmosphere is doing in places that are hard to get to. We need to know the temperature, humidity, air pressure, wind speed, and wind direction, not only at the surface but also in the upper atmosphere. We need information about what is happening on land and over the sea. There are literally thousands of people and pieces of automated equipment around the world filling this need for detailed climatic information. The World Meteorological Organisation (WMO), which operates under the auspices of the United Nations (UN) and contains 179 member countries, collects and distributes data from 10,000 land-based stations and 7,000 ship-based stations, from automatic weather sensors on buoys in the sea, attached to balloons, on aircraft, and from satellites.

Weather Balloons

Weather balloons filled with hydrogen gas are released from 1,000 sites in the United States every day at 7 am and 7 pm. The readings from the instruments are transmitted to the ground recording station, and the balloons are also tracked by radar to help measure wind speed and direction in the upper atmosphere. After about two hours, the average balloon reaches a height of 20 miles (32 km) and bursts. The instrument package parachutes back to the Earth's surface, where it is recovered and reused.

Weather Balloon

Ocean Buoy

Ocean Buoys

Ocean buoys deliver instrument readings about the conditions of the ocean, as well as measurements of standard air temperature, humidity, and pressure. Information about the surface of the ocean and the air above is essential when it comes to predicting the formation and pathway of hurricanes, cyclones, and typhoons. This information is also needed for computer models trying to predict the occurrence and intensity of El Niño and La Niña conditions.

Automated Weather Stations

Automated weather stations provide data, sometimes as often as every minute, and can record key weather data automatically 24 hours a day, non-stop. Although they are expensive, they make an invaluable contribution to the field of weather forecasting, as they provide vital information throughout the night when information is rarely collected by people-monitored weather stations.

Doppler Radar

A powerful remote sensing tool, Doppler radar is especially good at detecting thunderstorms that other instruments and forecasting methods find hard to detect. Conventional radar sends out short, powerful microwave pulses. A receiver picks up the signal, which has bounced back from any object encountered in the form of radio waves. In this case, it would be the precipitation particles in the clouds. Thus, radar provides a rough description of the object's size and estimated distance. Doppler radar measures the frequency difference between the signals bouncing off objects moving forward or away from the antenna. This allows scientists to build a picture of how the winds are moving within a thunderstorm, and thus predict the storm's future behavior.

Doppler Radar

Satellites

Just try to imagine what it was like trying to predict when large storms were going to arrive on the land from the sea in the days before satellites were invented. It was nearly impossible. The great advantage of satellites is that they provide a large picture of the weather at the time it is actually happening. Satellites can make pictures of the sunlight bouncing off clouds, they can use infrared radiation to track storms through the night, and other exotic instruments allow the satellites to measure the sea temperature, wind speed, and direction.

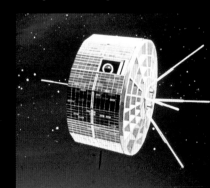

Satellite

CHAOS THEORY
Predicting the weather

The National Weather Service in the United States has recently spent US$4.5 billion on improving weather forecasting. Other countries are pouring similar resources into weather agencies. It's big business— predicting a storm can save billions of dollars and many lives.

IMPROVED FORECASTING

There is no doubt that weather forecasting is improving. Forecasts for the next three to four days are now as accurate as two-day forecasts were 15 years ago. Predictions of rain three days in advance are as accurate as one-day forecasts were in the mid-1980s. The accuracy of flash-flood forecasts has improved from 60 percent correct to 86 percent, and potential victims of these floods get 53 minutes' warning, compared with 8 minutes in 1986. The warning time before a tornado hits has increased from 5 minutes in 1986 to 12 minutes in 1998. Severe local thunderstorms and similar cloudbursts can be predicted 18 minutes beforehand, rather than 12 minutes over a decade ago. Seventy percent of all hurricane paths can be predicted up to 24 hours in advance, and hurricane landfall can be estimated to within 100 miles (160 km). But can we ever be 100 percent accurate?

PREDICTION AND UNPREDICTABILITY

These are great achievements, but it does not explain why, with all our technology and our understanding of the climate system, we cannot predict the weather farther ahead—say, ten days, a month, or even a year in advance. Moreover, think of all those times that the weather report has said it will be sunny and yet it rains. We all know they can get it wrong—but why does it happen?

In the 1950s and 1960s, it was thought that weather predictions were limited by our lack of data, and that if we could work with exact measurements and understand the fundamental processes, we would be able to achieve a more accurate level of prediction. Then, in 1961, meteorologist Edward Lorenz brought about a radical change in the way we think about natural systems: his discovery led to the development of Chaos Theory (see opposite). As a result, scientists now know that even the tiniest variations in atmosphere temperature, pressure, and humidity can have large and unpredictable effects on weather patterns.

We can predict (within certain boundaries) what the weather will be like, But when it comes to more detailed predictions, everything breaks down owing to what has become known as "the Butterfly Effect." The idea is that small changes, represented by the flapping of a butterfly's wings, can have a large effect on the weather (for example, altering the strength and direction of a hurricane), as errors and uncertainties multiply upward through the chain of turbulent features. We will never know which of the small weather changes will combine to have large effects.

Lorenz used twelve equations in his weather model—modern weather computers use 500,000. But even the best forecasts, which come from the European Centre for Medium Weather Forecasts based in Reading, United

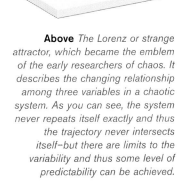

Above *The Lorenz or strange attractor, which became the emblem of the early researchers of chaos. It describes the changing relationship among three variables in a chaotic system. As you can see, the system never repeats itself exactly and thus the trajectory never intersects itself—but there are limits to the variability and thus some level of predictability can be achieved.*

Kingdom, are speculative beyond four-day predictions and worthless beyond a week—all due to Chaos.

Chaos Theory states that we can understand weather and predict general change. But it also shows us that even if we improve our weather monitoring, the chaotic effects of very small changes in weather prevent us from ever being able to predict exactly when and where storms such as hurricanes and tornadoes will occur.

GIANT LEAP FORWARD IN WEATHER PREDICTION

Weather prediction is set to take a massive leap forward. Up to now, computer models used for weather prediction have been limited both by the size of the area they deal with (global models use 37-mile [60-km] grid squares while local ones use 7-mile [12-km] grids) and the simplifications they make. For example, the models assume that the upward buoyancy of a body of air and the downward effect of gravity always balance out. This is a fair approximation of how big Atlantic Storms work, but it doesn't work for small-scale storms.

However, the United Kingdom's Meteorological Office is going to change everything. In a few years' time they will introduce forecasts down to 0.3 square miles (1 square kilometer). The new model will calculate how air moves vertically in storms, instead of modeling movement of air over a fixed point.

So, the experts might finally be able to predict rain at Wimbledon—a happy end to a long and wet tradition!

Top and above *The Mandelbrot Sets are the most visual of all the representations of chaos. Using simple mathematical equations, the mathematician Benoit Mandelbrot was able to create pictures which, no matter how much you enlarge them, are always as complicated and never repeat themselves. These fractal diagrams can also be seen in nature in snowflakes, each one of which is unique.*

EDWARD LORENZ– CHAOS AND SERENDIPITY

In 1960 Edward Lorenz, a meteorologist at Massachusetts Institute of Technology (MIT), produced one of the first computer models of weather patterns.

One day in the winter of 1961, Lorenz's computer model produced some very interesting patterns, which he wanted to look at in greater detail. He took a shortcut and started midway through the run. Of course, this was one of the earliest computers, so he had to retype all the starting numbers. Instead of typing them to six decimal places (0.506127), he typed only the first three (0.506) to save time and space, then went and made himself a cup of coffee.

When Lorenz returned, he found that the weather patterns had diverged so much from the initial run that there was no recognizable similarity between the two. It seems the model was very sensitive to the very small changes that Lorenz made—that one part in a thousand, instead of being inconsequential, had a huge effect on the outcome.

This ground-breaking work, discovered completely by chance, has led to the development of Chaos Theory.

PREDICTING HURRICANES

Detection, tracking, and landfall

We know that the general path taken by hurricanes is from east to west in the flow of the Trade Winds, both north and south of the equator. But predicting exactly where and when they will hit land is not always so easy: forecasts can be at least 100 miles (160 km) out. Why is this?

The simple answer is that hurricanes can be caught up in other wind circulations; ocean currents, too, can cause them to veer north or south before they eventually return to their general eastward direction. For example, in the North Atlantic Ocean, the high pressure associated with Bermuda and the Gulf Stream produces a northward-curving hurricane track.

Hurricane prediction is based on both oceanographic and atmospheric research. There are three key stages in hurricane prediction: detection, tracking, and landfall.

Top right *Three-dimensional image of Hurricane Linda, which hit in September 1997, reconstructed from satellite data.*

Right *Comparison between predicted hurricane paths and the actual path.*

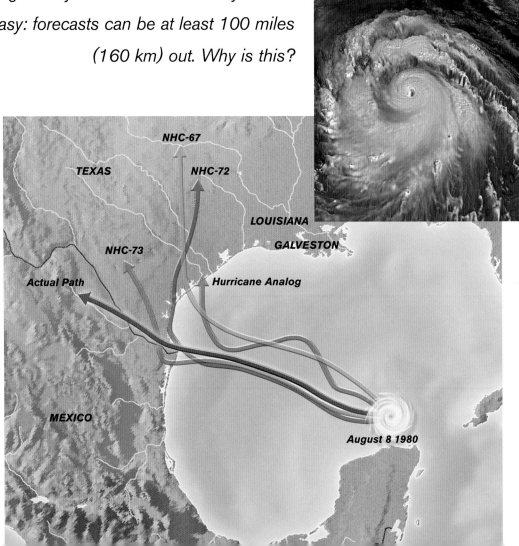

Top *National Oceanic and Atmospheric Administration research aircraft flying into the eye of Hurricane Caroline.*

Above *Computer-generated image of Hurricane Mitch in 1998 as it stalled off the coast of Honduras.*

DETECTION AND TRACKING

Weather satellites are the most important instruments for detecting and tracking hurricanes. They are used to detect clusters of rain clouds, which are in turn tracked to see if they evolve into a hurricane. Two types of satellite are used: polar orbiting satellites, which spin around the Earth from pole to pole, and geostationary satellites, which maintain the same position above Earth at all times. The sole drawback of satellite data is that it only shows the temperature and the cloud cover associated with the hurricane—it does not provide information about the processes going on within the storm, which is essential to understanding how the hurricane will develop.

PREDICTION OF LANDFALL

The most important hurricane prediction is when and where a hurricane will make landfall, as this enables local authorities to make provisions to evacuate people. Most hurricanes move according to well-defined paths, as future movement is directly related to the energy of the storm and its previous path. In 70 percent of hurricanes, the paths can be predicted up to 24 hours in advance, based on their direction and speed for the previous 36 hours. However, this still leaves 30 percent of hurricanes that cannot be predicted simply. Of these, nine out of ten can be predicted using a combination of very detailed climate data and large complex computer models. However, even the complex computer models do not always get it right: Hurricane Mitch is a prime example (see pages 60–63).

Despite the fact that most hurricanes have predictable paths, estimating the landfall is very difficult, as hurricanes do not move at a steady pace or in straight lines. They stop and start, making a zig-zag along their path. Hence the margin of error involved in predicting when and where a hurricane will land can be at least 100 miles (160 km).

Nonetheless, good progress has been made in the prediction of landfall, and every year the error drops by one percent. This may not sound like much, but this has meant that since 1970, the error in estimating a hurricane landfall in a 24-hour forecast has fallen from 140 miles (225 km) to 100 miles (160 km).

WARNING

Because of this inherent margin of error, forecasters try to delay issuing hurricane warnings for as long as possible, in order to prevent expensive preparations that may not be required. Therefore, specific hurricane warnings are usually given no more than 12 to 18 hours in advance of landfall. In developed countries such as the United States, this is ample time to implement evacuation plans to prevent loss of life. However, in poorer developing countries, where communications are not very good, this is not enough time.

RELIEF AND RESCUE
Rebuilding lives

Storms and other natural disasters will always occur. With more research, scientists will be able to better predict when and where killer storms will hit. However, governments, rescue agencies, and charities know that they must plan for the worst.

WHERE TO START?

Right is the sequence of rescue events that should follow a storm impact. The most important thing for the local authorities to do once the initial damage is assessed is to organize the immediate rescue of trapped people, which includes:

- People within a collapsed building. Note that after only six hours, less than half the people trapped under rubble will still be alive.
- People buried in avalanches. Most die within 30 minutes if they are not found.
- People buried by mudslides brought on by flooding. For example 1,200 people were killed in a mudslide on the side of the Casitas Volcano in Nicaragua in the aftermath of Hurricane Mitch.
- People trapped by floods who are out of the water but are stuck up isolated trees or on the tops of buildings.

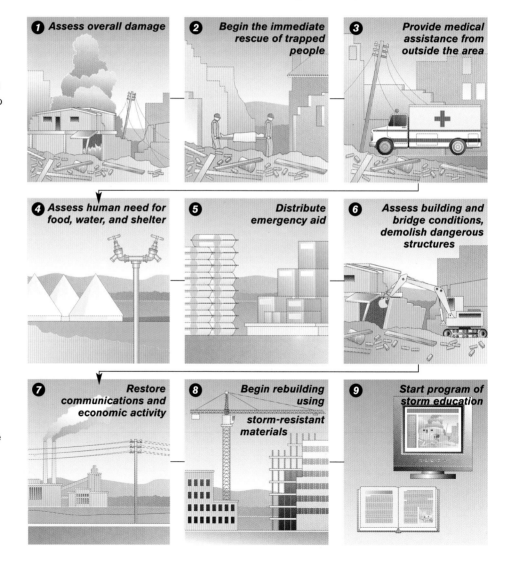

1. Assess overall damage
2. Begin the immediate rescue of trapped people
3. Provide medical assistance from outside the area
4. Assess human need for food, water, and shelter
5. Distribute emergency aid
6. Assess building and bridge conditions, demolish dangerous structures
7. Restore communications and economic activity
8. Begin rebuilding using storm-resistant materials
9. Start program of storm education

Right International rescue teams arriving in Tegucigalpa, Honduras, to help with the aftermath of Hurricane Mitch.

Below right Rescue teams with sniffer dogs were used in Tegucigalpa to find survivors in the collapsed buildings and mudslides.

STORMS DESTROY THE INFRASTRUCTURE

The major problem with storm disasters is that rescue and relief is usually attempted while the effects of the storms are still being felt. For example, the floodwaters in Central America after Hurricane Mitch took over ten days to recede. Prolonged storm conditions also affect the distribution of medicines, food, water, and the provision of shelter. With nearly all the roads and bridges destroyed in Honduras, plus the shortage of helicopters, it was impossible to get aid to the people who needed it most.

DISEASE

The damage done by storms does not stop with the initial loss of life. Although studies have shown that the majority of deaths occur within the first two days of a storm and associated floods, disease can follow. Many of the worst diseases have a four- to five-day incubation period and thus, if they are not controlled, can kill as many people as the storm itself.

The authorities must control and monitor disease, because simple everyday things such as fresh water,

food, and drainage may have been disrupted by the storm or flood.

Sanitation must be made a matter of priority. If the possible sources of disease (for example, contaminated water) are controlled, then the number of deaths can be dramatically reduced. In 1998, the floods in Bangladesh lasted two months and covered three-quarters of the country, and there was a major outbreak of cholera and other waterborne diseases. The spread of disease was also particularly acute after the devastation of Hurricane Mitch in Central America. With so many unburied dead bodies, so much stagnant water from the flood, and continued inaccessibility to many communities, many diseases were able to proliferate and take their toll on those who had survived the hurricane.

Above left *Dogs are essential when searching through an avalanche as they can detect people buried deep under the snow.*

Left *Flooding in Mozambique in 2000 was so extensive that the only way to rescue people was by helicopter.*

Right *In the town of Redon in France in 2000, the floods were so bad that elderly people had to be evacuated from their homes by boat.*

Left *After a disaster such as Hurricane Mitch, it is essential to provide the victims with fresh, clean drinking water as disease can sometimes kill as many people as the initial storm or flood.*

Above *International relief aid is essential in the aftermath of a regional disaster because those areas which could have helped with food and clean water are themselves severely affected.*

TRIAGE

Once people have been rescued, the usual principle of "first-come, first-served" is inadequate, considering the limited medical resources available during the aftermath of a serious storm. Hence a system called triage is used. One type of triage system, whereby patients are seen depending on the severity of the injury, with the most life-threatening illnesses and injuries attended to first, is used in your local emergency room. During a disaster, however, triage is very different and consists of rapidly classifying patients (usually with a tag) to identify those who would gain the most from immediate medical attention. This is not related to the seriousness of the injury. The highest priority is given to those patients who, with simple medical care, will dramatically improve their short-term or long-term health. The lowest priority is given to patients who are nearly dead—those who, even with intense medical care, may not survive—and to the slightly injured who can wait for assistance without coming to further harm. This is because the best use must be made of very limited medical resources: during the most destructive storms, hospitals, electricity, and transportation are all wiped out.

PREPARATION AND MITIGATION

Money is the key

We know that we cannot avoid natural disasters, but we need to ensure we are well prepared for them. Educating as many people as possible in evacuation procedures is a good first step. In the longer term, we need to think about local building conventions and town planning. As ever, the main obstacle is a lack of money.

Below left *Victims of storm disasters need shelter. Tents are an excellent starting point and buy time for the local authorities to sort out more solid housing.*

Below *We can take measures to protect ourselves: for example, barriers can shield villages and towns from the worst effects of avalanches.*

The immediate need for housing after a storm or flood may involve accommodating survivors in hotels, mobile trailers, holiday homes, public buildings, even on ships, trains, and buses.

On a simpler scale, plastic sheeting can be used; it is cheap, easy to transport, and protects against the elements. Tents can also be very useful, and are often stockpiled by local authorities.

In the longer term, in developed countries where transportation networks are good, prefabricated buildings are a great solution to housing needs. A 430 square foot (40 m^2) prefabricated building can be easily transported on a flatbed truck. The other advantage is that these buildings can be connected to water, gas, electricity, and sewage. As soon as people are housed, redevelopment can start.

In less developed countries the problems are different. On average only one-fifth of the cost of redevelopment comes from foreign aid—so the key to emergency housing in these countries is to keep the solutions cheap and simple. Relief agencies have started teaching different rapid-housing techniques—for example, stack-sack. Stack-sack houses are constructed from sandbags filled

with cement mix. Steel rods are placed through the sandbags before the cement sets so as to support and strengthen each wall.

EDUCATION

As with any natural disaster, damage and loss of life from storms can be mitigated but never entirely prevented. The best form of mitigation is to start a program of storm education. An important

element of such a program is to assess the risk of storms and to plan urban areas so as to minimize damage in future disasters (for example, by strengthening the existing structures and introducing strict building laws, as was done in Jamaica after Hurricane Gilbert hit in 1988).

Storm education should also involve teaching evacuation procedures and informing people how they should react in the event of a major storm.

THE PRICE OF POVERTY

All these measures to minimize the risks and to educate sound very sensible but there is one major problem–they require money. It is unfortunate that the Central American countries affected by hurricanes are already extremely poor and are amongst those countries most crippled by international debt.

This debt was originally borrowed in order to try and help the country develop, so it is paradoxical that repaying it does just the opposite. It is common for less developed countries to spend

Above *After every storm or flood disaster, lessons must be learned, because when an area is hit it is rarely the first time such an event has occurred. Thus everything from stricter building laws to evacuation procedures needs to be examined if damage from these storm events is to be mitigated in the future.*

Left *Starving people in Ethiopia wait at a distribution center for food supplied by international aid agencies.*

Below left *The poverty is so bad in Mozambique that during the floods in 2000 they did not have enough helicopters or boats to rescue people or distribute food. South Africa mobilized its army and air force to help with the relief and rescue.*

Below right *Poverty is made worse by the amounts of money that many less developed countries spend on weapons.*

twice as much paying back their debt as they do on basic health and education combined.

In Central America, for example, Hurricane Mitch destroyed the infrastructure, isolated hundreds of communities, and allowed starvation and disease to become rampant. It took weeks for the relief needs to be assessed and for aid to be distributed to those who needed it. Why? One of the main reasons was that all available money was tied up in the affected countries' international debt repayments.

Many people around the world are pushing for these international debts to be cancelled. It simply does not make sense to burden countries with an immense debt that they cannot realistically ever hope to repay.

The table opposite shows the money owed by each of the four nations worst hit by Hurricane Mitch. The international debts of these countries represent between 15 and 310 percent of the annual income before the hurricane hit—a colossally high figure.

SOCIO-ECONOMIC CONDITIONS IN THE FOUR COUNTRIES WORST HIT BY HURRICANE MITCH IN 1998

Honduras

Population:	5.8 million
Number of people killed by Hurricane Mitch:	more than 11,000
Number of people made homeless by Hurricane Mitch:	up to 2 million
International Debt:	US$4,453 million
Gross National Product (GNP) per capita:	US$600
Inflation:	25.4%
Life expectancy:	68.8 years
Access to safe water:	87%
One doctor every:	1,300
Under-fives deaths:	54 per 1,000
Illiteracy:	27%

El Salvador

Population:	5.6 million
Number of people killed by Hurricane Mitch:	more than 350
Number of people made homeless by Hurricane Mitch:	at least 100,000
International Debt:	US$2,894 million
Gross National Product (GNP) per capita:	US$1,610
Inflation:	15.5%
Life expectancy:	69.4 years
Access to safe water:	55%
One doctor every:	1,563
Under-fives deaths:	56 per 1,000
Illiteracy:	29%

Nicaragua

Population:	4.1 million
Number of people killed by Hurricane Mitch:	more than 4,000
Number of people made homeless by Hurricane Mitch:	unknown
International Debt:	US$5,929 million
Gross National Product (GNP) per capita:	US$380
Inflation:	8.6%
Life expectancy:	67.5 years
Access to safe water:	58%
One doctor every:	2,000
Under-fives deaths:	68 per 1,000
Illiteracy:	34%

Guatemala

Population:	10.9 million
Number of people killed by Hurricane Mitch:	more than 200
Number of people made homeless by Hurricane Mitch:	at least 80,000
International Debt:	US$3,785 million
Gross National Product (GNP) per capita:	US$1,340
Inflation:	19.5%
Life expectancy:	66.1 years
Access to safe water:	62%
One doctor every:	4,000
Under-fives deaths:	70 per 1,000
Illiteracy:	44%

MILITARIZATION EXACERBATES POVERTY

Yet another problem for the less developed countries is the fact that they are often ruled by a small elite who spend a lot of money on weapons to protect their power. This militarization helps maintain the status of the very few at the top of the hierarchy, increasing the poverty of the majority, and reducing investment in programs aimed at mitigating the damage caused by natural disasters. This in turn can lead to political instability, which leads to increased military spending in order to protect the power of the elite. A vicious cycle is established.

Mozambique is a good example of this. In 1990 Mozambique spent over 10 percent of its gross national product (GNP) on the military, and less than 4 percent on health. Although by 1999 military spending had dropped to just over 2 percent, financially Mozambique was extremely vulnerable when massive floods hit in 2000.

Above *The table shows the poverty of the four countries most affected by Hurricane Mitch in 1998. Note the crippling international debts which each country has. Many people around the world are trying to get these debts cancelled so that these countries can start to develop.*

CHAPTER 3

The Future

We have already seen that Earth is a stormy and turbulent planet. As temperatures rise, global warming will put even more energy into the system. The growing temperature difference between the poles and the equator will lead to more frequent and more ferocious storms—more powerful than any ever recorded in history.

What other surprises might be in store? Warmer temperatures could evaporate the stores of gas hydrates that lie beneath the oceans, releasing massive quantities of methane into the atmosphere and accelerating global warming still further. The melting of the world's ice sheets could leave so much freshwater on the oceans' surface that the deep-water circulation upon which our weather depends would simply be turned off. One thing is certain: the 21st century will be one of huge climatic instability.

Above *Ice floes like these in Antarctica might disappear within a hundred years if global warming continues at its present pace.*

GLOBAL WARMING
The greenhouse effect

The temperature of the Earth is controlled by a delicate balance between the input of the Sun's energy and the loss of this energy back into space. The so-called greenhouse gases are critical to this temperature balance, but changes in the relative proportions of these gases are contributing to climate change the world over.

THE EARTH'S NATURAL GREENHOUSE

The energy that the Earth receives from the Sun is in the form of short-wave energy, or radiation. On average, about one-third of the solar radiation that hits the Earth is reflected back into space while most of the remainder is absorbed by the land and oceans. The energy warms the Earth's surface which, as a result, emits heat, or long-wave infrared radiation. Gases in the troposphere (see pages 18–19), known as greenhouse gases, can trap some of this long-wave radiation, thus warming the atmosphere.

Naturally occurring greenhouse gases include water vapor, carbon dioxide, ozone, methane, and nitrous oxide. Together, they create a natural greenhouse (or blanket) effect, warming the Earth by 63°F (35°C). Although the greenhouse gases are depicted in the figure below as a layer, this is only to show their blanket effect, as they are, in fact, mixed throughout the atmosphere.

GREENHOUSE GASES AND CLIMATE

A planet's climate is determined by its mass, its distance from the Sun, and the composition of its atmosphere—in particular, the amount of greenhouse gases. One way to understand the Earth's natural greenhouse is to compare it with its two nearest neighbors.

Mars is smaller than Earth, so its gravity is less and hence it can retain only a small atmosphere. Its atmosphere is about 100 times thinner than that of the Earth and consists mainly of carbon dioxide. Mars' average surface temperature is about -58°F (-50°C), so most of Mars' carbon dioxide is frozen in the ground. Venus has almost the

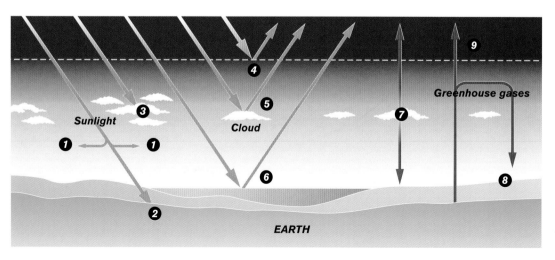

Sunlight

Cloud

Greenhouse gases

EARTH

Left *This diagram shows both the natural energy balance and the greenhouse effect. (1) Solar radiation or sunlight enters the atmosphere and is absorbed in various ways: by water vapor and gases, (2) by the Earth, (3) by clouds, (4) reflected back into space by the top of the atmosphere, (5) reflected back into space by the clouds or (6) reflected back into space by the surface of the Earth. Once the sunlight has been absorbed it is re-emitted as infrared radiation by (7) the clouds or (8) the Earth. (9) Some of this infrared radiation is captured by greenhouse gases, thus warming the atmosphere, and some of it is lost into space.*

same mass as Earth but a much thicker atmosphere, which is composed of 96 percent carbon dioxide. So Venus has a surface temperature of 860° F (460°C).

The Earth's atmosphere is very different. Our atmosphere is composed of 78 percent nitrogen, 21 percent oxygen, and 1 percent other gases. These other gases include the greenhouse gases.

The two most important greenhouse gases are carbon dioxide and water vapor. Currently, carbon dioxide accounts for just 0.03–0.04 percent of the atmosphere, while water vapor varies from 0–2 percent. Without the greenhouse gases trapping some of the radiation that is being reflected back into space, Earth's average temperature would be roughly -4˚F (-20°C).

However, the Earth's climate is unstable compared with that of Mars and Venus, and the relative proportions of greenhouse gases in the atmosphere can change very easily. Remember the Chaos Theory (pages 74–75)? Small changes can have a huge knock-on effect.

Human activities such as industrial processes and traffic pollution, which emit carbon dioxide into the atmosphere, are disturbing the natural balance, causing greenhouse gas levels in the atmosphere to increase. Most scientists believe this will lead to global warming–and temperature is one of the key factors in climate.

RESEARCH

The Intergovernmental Panel on Climate Change (IPCC), the organization of international scientists charged with investigating global warming, has identified the main greenhouse gases, where they come from, and their warming potential. The warming potential is calculated assuming that carbon dioxide has a potential of one. There are other, more dangerous, greenhouse gases but luckily they are found in very low concentrations.

Top right Global warming is caused by all the extra carbon dioxide, methane, nitrous oxides and CFCs that are being pumped into the atmosphere due to industrial processes and traffic.

Right Temperature and carbon dioxide concentration in the atmosphere over the past 400,000 years (from the Vostok Ice Core).

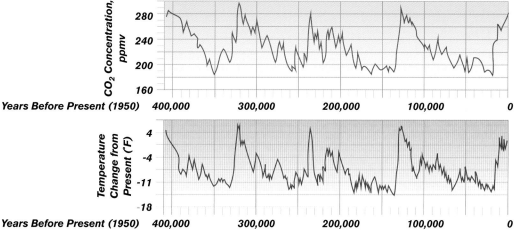

SHOULD WE BE WORRIED ABOUT GLOBAL WARMING?

Yes, and the reason why we should is that there is clear proof that atmospheric carbon dioxide levels have been rising since the beginning of the industrial revolution. The first measurements of carbon dioxide concentrations of the atmosphere started in 1958, at an altitude of about 13,120 feet (4,000 m) on the peak of Mauna Loa in Hawaii. The measurements were made here because the area is remote from local sources of pollution. What they have clearly shown is that atmospheric concentrations of carbon dioxide have increased every year since 1958. The mean concentration of approximately 316 parts per million by volume (ppmv) in 1958 rose to approximately 369 ppmv in 1998. The annual variation seen in the graph is due to carbon dioxide uptake by growing plants. The uptake is highest in the Northern Hemisphere springtime–hence every spring, there is a drop in atmospheric carbon dioxide, which unfortunately does nothing to the overall trend towards ever-higher values. This carbon dioxide concentration data from Mauna Loa can be combined with the detailed work on ice cores to produce a complete record of atmospheric carbon dioxide since the beginning of the industrial

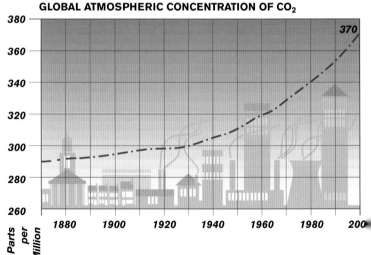

GLOBAL ATMOSPHERIC CONCENTRATION OF CO$_2$

Parts per Million

Left *Cars are a major source of greenhouse gases. Both the Indian and Chinese governments want every family to have one car by the year 2050. This would mean an extra billion cars in the world, all pumping out pollution.*

Above *Records from ice cores show that levels of global atmospheric carbon dioixde have been increasing over the last 120 years.*

revolution. What this shows is that atmospheric carbon dioxide has increased from a pre-industrial concentration of about 280 ppmv to over 370 ppmv at present–an increase of 160 billion tonnes, representing a 30 percent increase overall.

To put this increase into context, we can look at the changes between the last ice age, when temperatures were much lower, and the pre-industrial period. According to evidence from ice cores, ice-age atmospheric carbon dioxide levels were about 200 ppmv, compared to pre-industrial levels of 280 ppmv. This carbon dioxide increase was accompanied by a global warming of 9°F (5°C) as the world freed itself from the grips of the ice age. Admittedly, there were many other causes for

the end of the ice age and the subsequent warming, but carbon dioxide did play an important role. What these figures demonstrate is that the level of pollution we have already caused is comparable to natural variations, which took thousands of years.

CO₂ CONCENTRATION IN THE ATMOSPHERE: MAUNA LOA CURVE

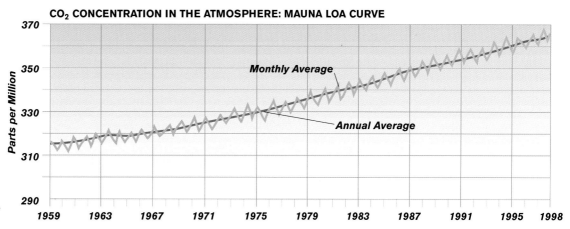

Monthly Average

Annual Average

Top left *Coal-fired power stations produce both carbon dioxide, which adds to global warming, and aerosols, which add to local cooling.*

Top right *Pollution belches out of all our factories 24 hours a day*

Above *The concentration of carbon dioxide in the atmosphere has been measured at Mauna Loa in Hawaii. The annual variation seen in the graph is because of carbon dioxide uptake in plants which is greatest during spring in the Northern Hemisphere. Unfortunately, this does not stop the overall trend toward ever-higher values.*

THE POLLUTION PRODUCERS

Only we can make the difference

In July 2001, the leaders of the world met in Bonn, Germany, and produced the first international agreement on reducing global carbon dioxide emissions. However, this agreement has been made without the inclusion of the United States, and with very small reduction targets.

Fossil fuel emissions are not evenly distributed around the world. A significant part of carbon dioxide emissions comes from energy production, industrial processes, and transport. As you can see, North America, Europe, and Asia emit over 90 percent of the global human-produced carbon dioxide. Moreover, they have historically emitted much more than less-developed countries, ever since the start of the industrial revolution in the latter half of the 18th century. Hence the agreement may affect the economies of the industrialized nations more than those of less-developed countries.

In addition to industry, changes in land use—mainly from the cutting-down of forests for agricultural use, urbanization, or roads—have also played a major part in increasing carbon dioxide emissions. When large areas of rainforest are cut down, for example, the land often turns into less productive grasslands, with

considerably less capacity to store carbon dioxide.

ECONOMIC DEVELOPMENT VS ECOLOGICAL DISASTER

So who is going to take responsibility? This is a major

source of debate. Non-industrialized countries are striving to increase their population's standard of living, thereby increasing the emission of greenhouse gases, since economic development is closely associated with energy production. Thus, the volume of carbon dioxide

Above *Cooling towers at a power station producing greenhouse gases.*

Below left and right *Carbon dioxide is produced by industrial processes and land-use changes. These, however, are not evenly distributed: North America, Europe, and Asia produce 95 percent of the industrial carbon dioxide, while South America, Africa, and Asia contribute 95 percent of the land-use emissions.*

CO_2 EMISSIONS FROM INDUSTRIAL PROCESSES (IN BILLION TONS)

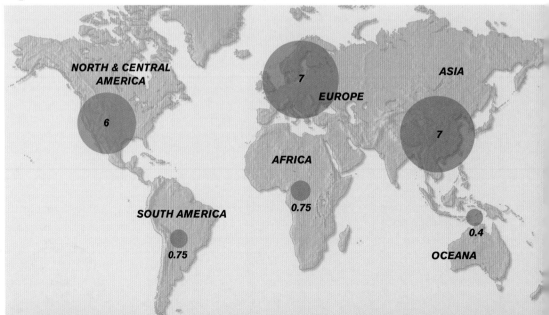

NORTH & CENTRAL AMERICA
6

EUROPE
7

ASIA
7

AFRICA
0.75

SOUTH AMERICA
0.75

OCEANA
0.4

will probably continue to increase—despite efforts to reduce emissions in the industrialized countries. China, for example, has the second-biggest emission of carbon dioxide in the world. However, per capita, the Chinese emissions are ten times lower than those of the United States, who top the list.

WORKING FOR CHANGE

The Intergovernmental Panel on Climate Change (IPCC) was jointly established in 1988 by the United Nations Environmental Panel (UNEP) and the World Meteorological Organization. (WMO) due to worries about the possibility of global warming. Its role is to continually assess the various aspects of climate change, including our scientific knowledge, the environmental and socio-economic impacts of climate change, and the best way to respond to the threats posed by global warming.

The IPCC is recognized as the most authoritative scientific and technical voice on climate change, and its assessments have had a profound influence on the negotiators of the United Nations Framework Convention on Climate Change (UNFCCC) and its Kyoto Protocol.

The meetings in The Hague in November 2000 and in Bonn in July 2001 were the second and third attempts to ratify the protocols that had been laid out in Kyoto in 1998. Unfortunately, President Bush pulled the United States out of the negotiations in March 2001. However, 186 other countries made history in July 2001, by agreeing the most far-reaching and comprehensive environmental treaty that the world has ever seen.

WHAT IS THE IPCC?

The IPCC is organized into three working groups, plus a task force to work out the amount of greenhouse gases produced by each country. Each of these four bodies has two co-chairs (one from a developed country and one from a developing country) and a technical support unit. Working Group I assesses the scientific aspects of the climate system and climate change. Working Group II addresses the vulnerability of human and natural systems to climate change, the negative and positive consequences of climate change, and options for adapting to them. Working Group III assesses options for limiting greenhouse gas emissions and otherwise mitigating climate change, as well as economic issues. The IPCC also provides governments with scientific, technical, and socioeconomic information relevant to evaluating the risks and developing a response to global climate change. Approximately 400 experts from some 120 countries are directly involved in drafting, revising, and finalizing the IPCC reports, and another 2,500 experts participate in the review process. The IPCC authors are always nominated by governments and international organizations, which include non-governmental organizations.

Below *Deforestation releases carbon dioxide into the atmosphere, contributing to global warming.*

CO$_2$ EMISSIONS FROM LAND USE CHANGE (IN BILLION TONS)

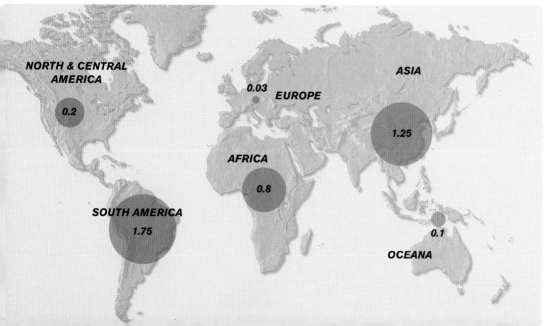

NORTH & CENTRAL AMERICA

0.2

0.03 EUROPE

ASIA

1.25

AFRICA

0.8

SOUTH AMERICA
1.75

0.1

OCEANA

THE EVIDENCE FOR CLIMATE CHANGE

Steady warming since the 1980s

The three main indicators of global warming are temperature, precipitation, and sea level. Data collated by the IPCC, in some cases looking at records that go back 100 years or more, show worrying trends in all three areas.

A VARIATION IN THE EARTH'S SURFACE TEMPERATURE SINCE 1860 (˚F)

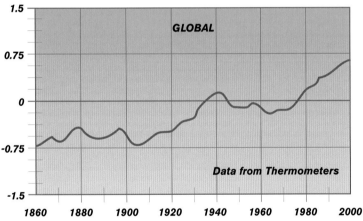

GLOBAL

Data from Thermometers

B VARIATION IN THE EARTH'S SURFACE TEMPERATURE IN THE LAST 1000 YEARS (˚F)

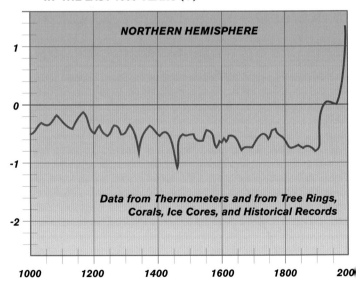

NORTHERN HEMISPHERE

Data from Thermometers and from Tree Rings, Corals, Ice Cores, and Historical Records

Above left and right (A) Average global surface temperatures since 1860, compiled from measurements of air temperature over the land and sea-surface temperature. A slight cooling can be seen in the 1970s, along with a dramatic rise in temperatures in the last two decades of the twentieth century. (B) When this is compared with the temperatures for the last 1,000 years, it is clear how much we have affected the global climate in the last century.

CHANGING TEMPERATURES

The mean global surface temperature has increased by about 0.5 to 1°F (0.3 to 0.6°C) since the late 19th century and by about 0.3 to 0.5°F (0.2 to 0.3°C) over the last 40 years–the period with the most reliable data. Recent years have been among the warmest since 1860, which is when records began.

This warming is evident in the temperature of both the sea surface and air over land masses, with indirect indicators, such as borehole temperatures and glacier shrinkage, providing independent support. It should also be noted that the warming has not been uniform across the globe. The recent warming has been greatest between 40°N and 70°N latitude, though some areas such as the North Atlantic Ocean have cooled in recent decades.

Some scientists suggest that some of this warming over the last 30 years is due to increased output of energy from the Sun.

However, this is not enough to explain the observed half-degree warming, and hence we have strong evidence that global warming is indeed caused by human activities.

CHANGING PRECIPITATION PATTERNS

Over the last ten years, the IPCC has put together all the precipitation records from around the world. Unfortunately, rainfall and snow are not as well documented as temperature, and the records do not go back as far. However, scientists have been able to put together general trends since the 1900s.

Averaged over the Earth's land surface, precipitation increased from the start of the 20th century up to about 1960, but has decreased since about 1980. It seems that precipitation has increased over land at high latitudes of the Northern Hemisphere, especially during the cold season—but decreases in precipitation occurred after the 1960s over the subtropics and the tropics from Africa to Indonesia. These changes are consistent with analyses of changes in stream flow, lake levels, and soil surface.

CHANGING SEA LEVELS

The IPCC has also put together data on sea levels which show that, over the last 100 years, the global sea level has risen by about 1.5 inches (4 cm) to 5.5 inches (14 cm). But sea-level change is

EVIDENCE FOR GLOBAL WARMING

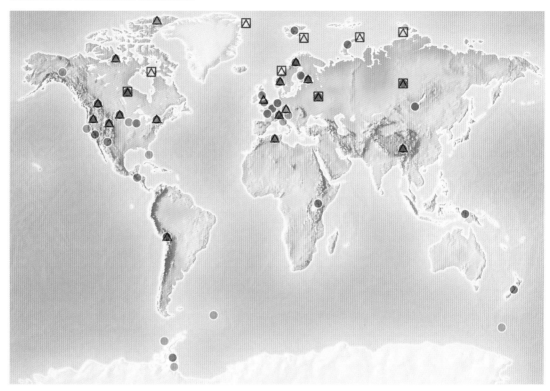

difficult to measure, and data has been derived mainly from tide-gauge measurements. In the conventional tide-gauge system, the sea level is measured relative to a land-based tide-gauge benchmark; however, the land surface is much more dynamic than one would expect and vertical movements are incorporated into the measurements.

However, it is likely that the observed rise in sea level over the last 100 years is related to the rise in global temperature over the same period. On this time scale, the warming and the consequent thermal expansion of the oceans may account for approximately

0.75 to 2.75 inches (2–7 cm) of the observed sea-level rise.

Other factors are more difficult to quantify. The rate of sea-level rise suggests that, while the retreat of glaciers and ice caps from the huge ice sheets of Greenland and Antarctica may account for about 0.75 to 2 inches (2–5 cm). However, lack of data on ice sheets over the last 100 years means that it is not possible to quantify their contribution to rising sea levels.

MELTING OF THE ICE CAPS

One of the biggest unknowns of global warming is whether the

KEY

AFFECTED...

⬤ *Hydrology and glaciers*
◐ *Sea ice*
⬤ *Animals*
⬤ *Plants*
△ *Studies covering large areas*
☐ *Studies using remote sensing*

Above *The most reliable evidence for global warming shows that it has already started to affect local climates.*

massive ice sheets that currently extend over Greenland and Antarctica will melt. One of the key indicators of the expansion or contraction of these ice sheets is the amount of sea ice that surrounds them. The state of the cryosphere (or the global ice) is extremely important, as shrinking ice cover causes the sea level to rise as the ice melts. Hence, in order to understand the effects of global warming on the cryosphere, we need to measure just how much ice is actually melting in the polar regions.

Unfortunately, submarines have already recorded a worrying thinning of the polar ice caps. Sea-ice draft–the thickness of the part of the ice that is submerged under the sea–was measured on submarine cruises in 1958, 1976, and between 1993 and 1997. Comparisons of these sets of data indicate that the mean ice thickness at the end of the melt season decreased by about 10 feet (3.1 m) in the 1990s. Ice draft

in the 1990s was over 3 feet (1 m) thinner than it had been two to four decades earlier. The main sea-ice draft had decreased from over 10 feet (3 m) to less than 6.5 feet (2 m) and the volume was down by some 40 percent.

In addition, in 2000 (for the first time in recorded history), a hole large enough to be seen from space opened in the sea ice above the north pole.

THE PERMAFROST UNFREEZES

Other evidence for global warming comes from the high-latitude and high-altitude areas, where it is so cold that for much of the year the ground is frozen solid to a great depth. This frozen ground is called permafrost, and during the summer months only the top 3 feet (1 m) or so gets warm enough to melt. This is called the active layer. Already in the permafrost of Alaska, there seems to have been a 5.5°F (3°C) warming down to at least 3 feet (1 m) over the last 50

years, showing that the active layer has become deeper.

With the massive increases in atmospheric carbon dioxide predicted for the future, it is likely that permafrost will increase in thickness, or in some areas, that the so-called discontinuous permafrost will disappear completely over the next century.

This widespread loss of permafrost will produce a huge range of problems in local areas because it will trigger erosion (or subsidence), change hydrologic processes, and release even more carbon dioxide and methane, trapped as organic matter in the frozen layers, into the atmosphere. Hence, changes in permafrost will reduce the stability of slopes and thus increase incidences of slides and avalanches. A more dynamic cryosphere will increase the natural hazards for people, structures, and communication links. Already buildings, roads, pipelines, and communication links are being threatened.

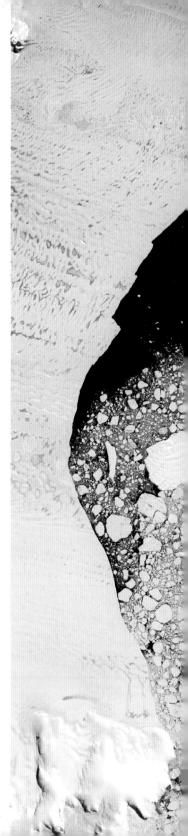

Left *The world's ice sheets contain enough water to raise sea level by 240 feet (75 m), but scientists still do not know how much of it will melt as a result of global warming.*

Right *A satellite image showing the initial break-up of the Larsen B Antarctic ice shelf, which completely collapsed in March 2002 sending 500 million billion tons of ice into the ocean. The question is whether this was a natural occurrence or due to global warming.*

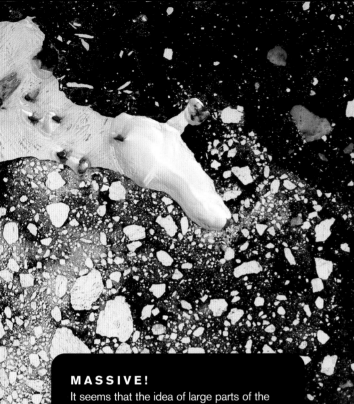

CHANGING WEATHER

There is lots of evidence that our weather patterns are changing: in recent years, for example, massive storms and subsequent floods hit China, Italy, England, Korea, Bangladesh, Venezuela, and Mozambique. English floods in 2000, classified as "once in 30-year events," occurred twice in the same month. Moreover the winter of 2000-01 was the wettest six months recorded in Britain since records began in the 18th century. In addition, on average, British birds nest two weeks earlier than 30 years ago. Insect species—including bees and termites—that need warm weather to survive are moving northward; some have already reached England by crossing the Channel from France. Glaciers in Europe are in retreat, particularly in the Alps and Iceland. Ice cover records from the Tornio River in Finland, which has been recorded since 1693, show that the spring thaw of the frozen river now occurs a month earlier.

There is also evidence for more storms in the North Atlantic. Wave height in the North Atlantic Ocean has been monitored since the early 1950s, from lightships, ocean weather stations, and, more recently, satellites. Between the 1950s and 1990s, the average wave height has increased from 8 feet (2.5 m) to 11.5 feet (3.5 m), an increase of 40 percent. Storm intensity is the major control on wave height, so we have clear evidence for a major surge in storm activity over the last 40 years. This was confirmed by a fantastic piece of science. German scientists found that storm-generated ocean waves pounding the coasts of Europe produce long-wave vibrations, which are picked up by the sensitive equipment set up to record earthquakes. From this evidence, they calculated the number of storm days per month during the winter. It seems that over the last 50 years these have increased from 7 to 14 days per month.

Above *Flooding seems to be becoming more common, with both England (top) and Bangladesh (above) being hit by massive floods in the last few years.*

MASSIVE!

It seems that the idea of large parts of the Antarctic ice shelf collapsing due to global warming is no longer science fiction. In March 2002, the Larsen B ice shelf completely disintegrated in less than 31 days. This ice shelf was 650 feet (200 m) thick and covered 1,255 square miles (3,250 square km). It sent 500 million billion tons of ice into the Weddell Sea.

This collapse dumped more ice into the ocean than all the collapses in the last 30 years put together. It also closely follows the massive iceberg, called B22, which broke off the Thwaites ice tongue in early March 2002. B22 is more than 40 miles (64 km) wide, 53 miles (85 km) long and covers an area of 2,124 square miles (5,500 square km).

Glaciologists have found that the Antarctic peninsula has warmed by 4.5°F (2.5°C) over the last 50 years and it seems to be extremely sensitive to global warming. The theory that enough meltwater could enter the Southern Ocean to switch off Antarctic Bottom Water formation may not, therefore, be too far-fetched!

MODELING THE FUTURE

Difficulties and variables

Future changes to the climate could have far-reaching effects. However, we do not know exactly what lies in store, as we are still unable to construct accurate predictive models.

Below *The present carbon cycle is complex: not only are there stores of carbon, but exchanges of carbon between the Earth and the atmosphere occur all the time. In the illustration below, stores and yearly exchanges of carbon are given in gigatons (1,000 million tons).*

THE CARBON CYCLE

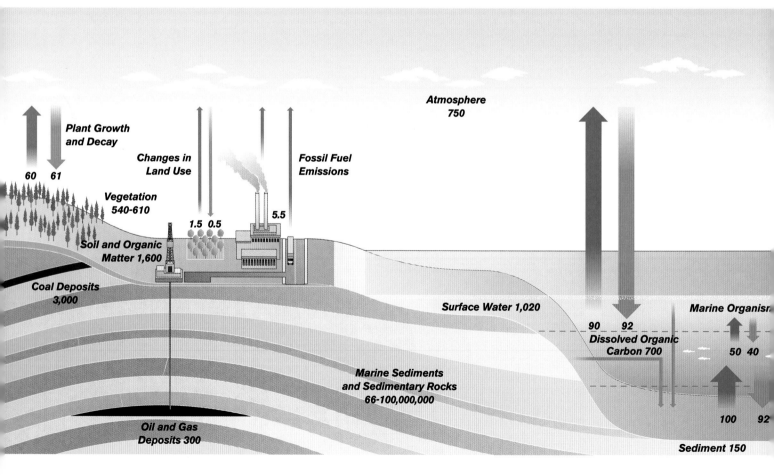

Plant Growth and Decay
60   61

Changes in Land Use

Fossil Fuel Emissions

Vegetation 540-610

1.5 0.5

5.5

Soil and Organic Matter 1,600

Coal Deposits 3,000

Oil and Gas Deposits 300

Atmosphere 750

Surface Water 1,020

Marine Sediments and Sedimentary Rocks 66-100,000,000

90   92

Dissolved Organic Carbon 700

Marine Organisr

50 40

100   92

Sediment 150

COOLING FACTORS

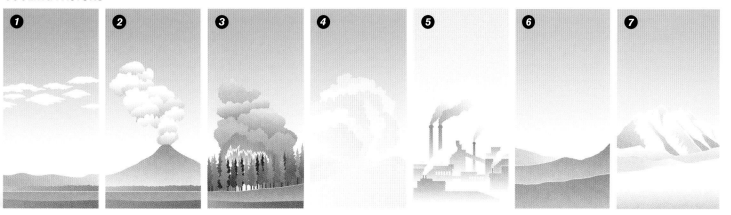

WHY IS IT SO DIFFICULT TO MODEL THE FUTURE?

To be able to predict the effects of global warming we need to understand the present-day carbon cycle. The Earth's carbon cycle is extremely complicated, with both sources and sinks of carbon dioxide. The illustration opposite shows fluxes in the global carbon reservoirs in GtC (gigatons, or 1,000 million tons). These figures are annual averages over the period from 1980 to 1989, and it must be remembered that what is shown here is a simplified version. The carbon system is, in fact, linked to the climate system. Moreover, there is evidence that many of the fluxes can vary significantly from year to year.

The surface ocean takes up just less than half the carbon dioxide produced by industry per year. However, there is still considerable debate about whether the oceans will continue to be such a large sink of our pollution.

COOLING EFFECTS

An added complication is that there are also, of course, cooling effects. This includes the amount of aerosols in the air, many of which come from human pollution—for example, sulfur emissions from power stations. These aerosols have a direct effect on the amount of solar radiation hitting the Earth's surface, and may have significant local or regional impact on temperature. In fact, the United Kingdom Meteorology Office has figured them into their computer simulations of global warming, and they provide an explanation of why industrial areas of the planet have not warmed as much as previously predicted.

Water vapor is a greenhouse gas, but at the same time, the upper white surface of clouds reflects solar radiation back into space. This reflection is called albedo—and clouds and ice have a high albedo and so reflect large quantities of solar radiation from surfaces on Earth.

Predicting what will happen to the amount and types of clouds, and the extent of global ice in the future, creates huge difficulties in calculating the exact effect of global warming. Open water absorbs heat, while white ice and snow reflect it, so if the polar ice caps were to melt, the albedo would be significantly reduced. This would be a positive result of the effects of global warming!

Another problem is that we do not really understand the role of clouds in the future. Some people argue that increased global warming will increase evaporation and thus form clouds, which will help to reflect more sunlight and keep the planet cool. Other scientists argue that a warmer Earth will cause clouds to rise higher in the atmosphere, where they are less effective at reflecting sunlight, so this will help to warm the planet. Most models assume this effect will balance out, but it is still one of the great uncertainties in our predictions of the future.

Above *Factors which cool the climate due to increased reflection of sunlight include (**1**) white tops of clouds, (**2**) aerosols produced by volcanic eruptions, (**3**) dust from forest fires, (**4**) dust from deserts, (**5**) aerosols from industrial processes, (**6**) barren land, and (**7**) ice- and snow-covered land that has a very high albedo.*

FUTURE CARBON DIOXIDE EMISSIONS

It is difficult to estimate just how much carbon dioxide will be emitted in the future, as there are so many unknown factors. Emissions will be influenced by population growth, economic growth, fossil fuel usage, the deforestation rate, and whether an international agreement to cut emissions is ever reached.

The Intergovernmental Panel on Climate Change (IPCC) has made great advancements in this area by publishing a region-by-region assessment of climate change on the United States. It has produced a model for the worst- and best-case future scenarios. The worst-

KEY

+	Large increase
+	Small increase
0	No change
-	Small decrease
-	Large decrease
I	Inconsistent sign

Right *Predictions of precipitation are much harder to produce. This map shows the general agreement between the computer models about where precipitation will increase and where it will decrease in both summer and winter over the next hundred years.*

Below left and right *Collation by the IPCC of computer-model predictions of future warming of the Earth. All the models show a significant increase in the global mean temperature by 2100, but there is huge uncertainty about exactly how great the increase will be.*

FUTURE PRECIPITATION PREDICTIONS

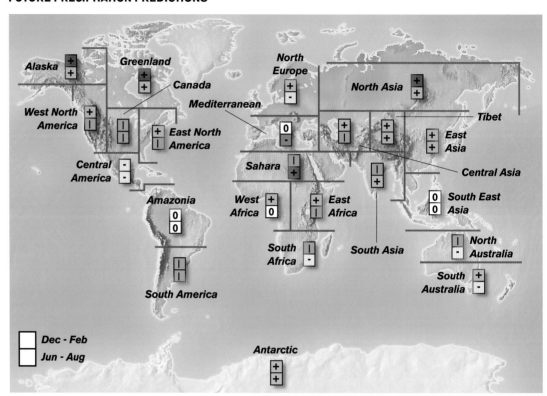

Dec - Feb
Jun - Aug

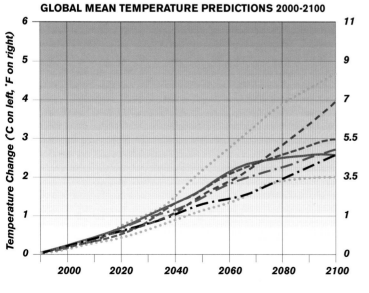

GLOBAL MEAN TEMPERATURE PREDICTIONS 2000-2100

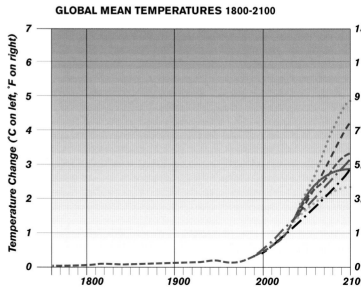

GLOBAL MEAN TEMPERATURES 1800-2100

ARCTIC SEA-ICE THICKNESS 1950's TO 2050's

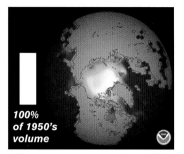

**100%
of 1950's
volume**

10-year average, center = 1955

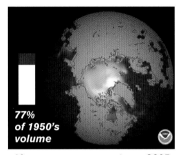

**77%
of 1950's
volume**

10-year average, center = 2005

**47%
of 1950's
volume**

10-year average, center = 2055

0 100 200 300 400 500

Thickness in cm

Left *The reduction in Arctic ice since the 1950s, simulated in this computer model, is consistent with observational estimates. Dramatic reductions are projected to continue for the next 50 years and beyond.*

case scenario shows an increase of 220 percent in atmospheric carbon dioxide, compared with pre-industrial levels, and the best-case scenario shows a 75 percent increase. Even if human-made carbon dioxide emissions are stabilized or reduced, the carbon dioxide content in the atmosphere will still rise for up to 100 years.

FUTURE GLOBAL TEMPERATURES

The IPCC has calculated the global mean temperature changes between 1990 and 2100. These climate models show that the global mean surface temperature could rise by about 2.5°F (1.4°C) to 10.5°F (5.8°C) by 2100. The topmost curve assumes constant aerosol concentrations beyond 1990 and high climate sensitivity and shows an increase of 2.5°F (1.4°C) by 2100. The lowest curve also assumes constant aerosol concentrations beyond 1990, but a much lower climate sensitivity and an increase of 2.5°F (1.4°C). There is still about 6.1°F

(3.4°C) temperature difference in the most extreme estimates, which are based on how sensitive scientists believe the climate system to be!

FUTURE RISES IN SEA LEVEL

Again, using the different carbon dioxide emission scenarios, the IPCC has projected the average global sea level up to 2100. Although estimates of climate sensitivity, ice-melt parameters, and the emission levels vary, the models project an increase of between 7.75 inches (20 cm) and 35 inches (88 cm).

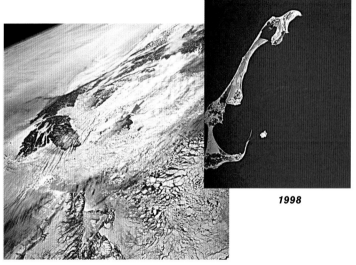

1998

1997

PREDICTED GLOBAL AVERAGE SEA LEVEL RISE

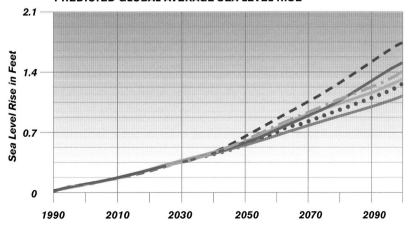

Above *The influence of global warming can be seen by the marked retreat of the ice which usually covers St Lawrence Seaway in winter. These two satellite pictures compare the winter of 1997 with that of 1998.*

Left *Collation by the IPCC of computer-model predictions of future sea-level rise. All models show a significant increase by 2100; the question is by how much.*

IMPACTS OF GLOBAL WARMING

Major climatic changes are ahead

Greenhouse gas emissions have already started to affect our climate. Future climate change will have massive impacts on the natural environment and human society, involving health, agriculture, water resources, forests, and wildlife.

SOCIO-ECONOMIC IMPACT

These changes will obviously affect the economies of the world. Although estimates have been made of the potential effect on individual socio-economic sectors, in reality the full consequences would be more complicated because impacts on one sector can also indirectly affect other sectors.

Below *The potential effects of global warming are terrifying, as they will affect both the natural environment and the world's economy. Below are the key areas in which global warming will have an impact.*

We've made a great deal of progress in understanding the climate system and climate change, but we must remember that projections of change and its impacts still contain many uncertainties. For a true assessment of the situation, we need to look at changes on a national and local level, as well as globally. Scientists are becoming more aware of this: the National Assessment Synthesis Team, for example, has just published its assessment of climate change in the United States, which deals with the impacts on a region-by-region basis.

ENVIRONMENTAL IMPACT OF CARBON DIOXIDE EMISSIONS

As we have seen, scientists have calculated that the global mean surface temperature could rise by between 2.5 and 10°F (1.4 and 5.8°C) by 2100. In addition, global mean sea level could rise by between 8 and 35 inches (20 and 88 cm) by 2100.

Their calulations are based on a number of different scenarios regarding carbon dioxide emissions. During the first half of the 21st century, carbon dioxide emissions will have relatively little effect on the projected sea-level rise, as most of the rise will be due to the large thermal inertia of the oceans. However, they will play an increasingly large role in the later part of this century, because of the effect of global warming on the ice sheets of both the Arctic and

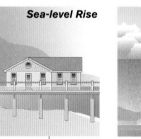
Sea-level Rise

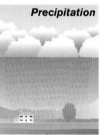
Precipitation

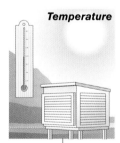
Temperature

IMPACTS ON...

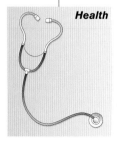
Health

Agriculture

Forest

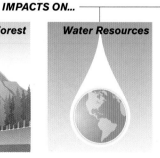
Water Resources

Coastal Areas

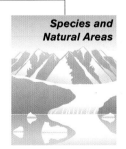
Species and Natural Areas

Antarctica. In addition, because of the thermal inertia of the oceans, sea level would continue to rise for many centuries beyond 2100, even if concentrations of greenhouse gases are stabilized.

Right This graph shows the change in temperature of the ground in Fairbanks, Alaska, at various depths. The top layer of soil is no longer frozen. If this melting moves deeper, it will start to affect the stability of buildings and roads.

Below There is strong evidence that Arctic sea ice is getting thinner. Here results from 1956 to 1976 are compared with data from the 1990s, showing a stark reduction.

PERMAFROST TEMPERATURES IN FAIRBANKS, ALASKA

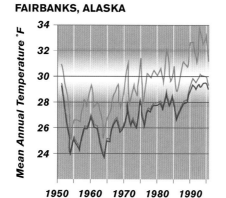

Soil depth in feet
········· 0.4   —— 1.7   —— 3.3

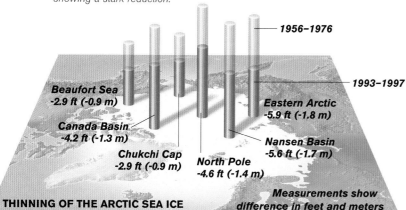

1956–1976

1993–1997

Beaufort Sea
-2.9 ft (-0.9 m)

Canada Basin
-4.2 ft (-1.3 m)

Chukchi Cap
-2.9 ft (-0.9 m)

North Pole
-4.6 ft (-1.4 m)

Eastern Arctic
-5.9 ft (-1.8 m)

Nansen Basin
-5.6 ft (-1.7 m)

THINNING OF THE ARCTIC SEA ICE

Measurements show difference in feet and meters

WHAT WE DON'T KNOW CAN HURT US...

What the sea-level calculation does not take into account is the possible melting of the world's ice sheets and glaciers. If all the ice sheets melted, their estimated contribution to sea-level rise would be as follows:
- Mountain glaciers = 1 foot (0.3 m)
- Western Antarctic ice sheet = 28 feet (8.5 m)
- Greenland = 23 feet (7 m)
- Eastern Antarctic ice sheet = 213 feet (65 m)

What is worrying is that NASA satellite measurements suggest that both Greenland and the Western Antarctic ice sheets are shrinking. If this produces enough meltwater, then we could have some big surprises in store in the future.

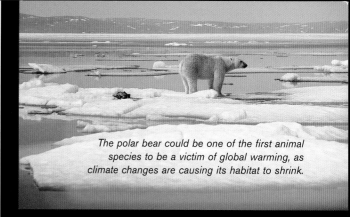

The polar bear could be one of the first animal species to be a victim of global warming, as climate changes are causing its habitat to shrink.

WATER
It has been predicted that by 2025 five billion people will be experiencing water shortages.

HEALTH
Diseases such as malaria, dengue fever, and cholera will affect tens of millions more people each year.

STORMS
Storms will become more common and stronger, and coastal areas will bear the brunt of this attack.

WILDLIFE AND NATURAL HABITATS
Climate change also has some terrible implications for wildlife and natural habitats. The species most at threat include:
- the mountain gorilla in Africa
- amphibians that only live in the cloud forest of the neotropics
- the spectacled bear of the Andes
- forest birds of Tanzania
- the Resplendent Quetzal in Central America
- the Bengal tiger and other species only found in the Sundarban wetlands
- polar bears and penguins
- rainfall-sensitive plants found only in the Cape Floral Kingdom of South Africa.

Natural habitats which are threatened include:
- coral reefs, mangroves, and other coastal wetlands
- montane ecosystems found in the upper 650–1,000 feet (200–300 m) of mountainous areas
- prairie wetlands
- permafrost ecosystems, and ice-edge ecosystems which provide a habitat for polar bears and penguins.

STORMS AND HURRICANES

Tempestuous times ahead

Rapid climate change leads to erratic weather patterns, and the number of storms increases. This happened in the UK during the Little Ice Age in the 17th century, although historical records prefer to remember the ice fairs that were held on the frozen River Thames.

Going into and coming out of the Little Ice Age produced some apocalyptic tempests. In 1703, for example, as climate was finally warming, over 8,000 people were killed in the worst storm ever recorded in British history.

INCREASE IN STORM ACTIVITY

It seems global warming may already be affecting global storms. There is good evidence of more storms in the North Atlantic Ocean during the second half of the 20th century. Wave height in the North Atlantic has been monitored since the early 1950s, from lightships, ocean weather stations, and, more recently, satellites. Between the 1950s and the 1990s, the average wave height increased by 40 percent, from 8 to 11.5 feet (2.5 to 3.5 m). Storm intensity is the major factor controlling wave height, so we have evidence of a surge in storm activity since 1960.

WHAT IT MEANS IN REAL TERMS

This increase in storm activity has resulted in a parallel increase in economic loss. Between 1951 and 1999, storms and floods were responsible for 76 percent of the global insured losses, 58 percent of the economic losses, and 52 percent of the fatalities from natural catastrophes. Property losses in the United States alone have increased tenfold since 1970. This is mainly because of the increased concentration of valuable property and infrastructure in low-lying coastal regions.

HURRICANES

For hurricanes and their cousins, cyclones and typhoons, to form in the tropics, the sea-surface temperature must at least 79°F (26°C) down to 200 feet (60 m) below the surface. All it then takes is a further increase of approximately 2°F (1°C) in sea-surface temperature to reduce atmospheric pressure enough to start the convective cell—a column

THE EVIDENCE

A fantastic piece of science confirmed the increase in storm activity. German scientists found that storm-generated ocean waves pounding the coasts of Europe produce long-wave vibrations, which are picked up by the sensitive equipment set up to record earthquakes. From this evidence they calculated the number of storm days per month during the winter. It seems that over the last 50 years these have increased from 7 to 14 days per month—more evidence that storms are becoming stronger and more frequent in the North Atlantic.

FINANCIAL LOSSES DUE TO STORMS AND FLOODS SINCE 1960

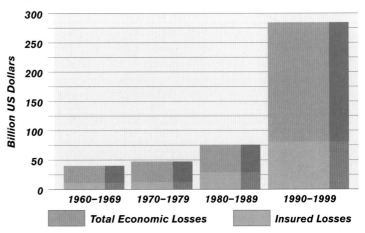

Billion US Dollars

| | Total Economic Losses | Insured Losses |

Left *Destructive hurricanes such as Chris could become more frequent in the future.*

of rapidly rising air, which sucks in air at sea level and produces the powerful hurricane vortex.

THE FUTURE

With increasing global warming, achieving the critical temperatures in the oceans will be easier than ever before, spawning more hurricanes with more energy to be unleashed upon our coastlines. Predictions by Dr. Mark Saunders, of the Benfield Greig Hazard Research Centre in London, suggest that by 2050 there could be as many as 50 percent more hurricanes hitting our coastlines. However, the good news is that there is currently no evidence to suggest that there has been an increase in the number of hurricanes or cyclones over the last 100 years.

This view of the future may become even more complicated, as some scientists argue that

Left *Losses from storm and flood damage around the world have increased dramatically since 1990, and global warming is set to exacerbate this situation.*

there will be an increased frequency of El Niño events. During El Niño there are fewer hurricanes in the Atlantic. This highly localized response to global warming makes it very difficult to predict the future.

If Dr. Saunders is correct, then there will be more numerous and powerful hurricanes, typhoons, and cyclones in our "Greenhouse World." The good news is that, because we know this, we should be able to improve evacuation procedures and consequently save lives.

HURRICANES AND THEIR IMPACT IN THE US, 1900–1995

From the graph on the right, you can see that although property losses due to hurricanes rose dramatically during the course of the 20th century, there was a correspondingly dramatic fall in the number of fatalities. This is due to improved storm forecasting and better evacuation procedures.

SEEDING HURRICANES

It has been suggested that "seeding," or dropping dry ice or silver iodide crystals into a hurricane could cause it to rain, thus reducing the hurricane's wind speed. However, there has been limited success with this: only in Hurricane Debbie in 1969 has the behavior of a hurricane been influenced by human interference. In the future we will need to that coastal areas will be increasingly vulnerable.

SEEDING HURRICANES

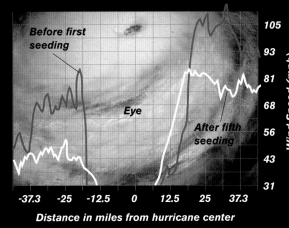

Before first seeding

Eye

After fifth seeding

Wind Speed (mph) at 12,000 feet altitude

105 · 93 · 81 · 68 · 56 · 43 · 31

-37.3 -25 -12.5 0 12.5 25 37.3

Distance in miles from hurricane center

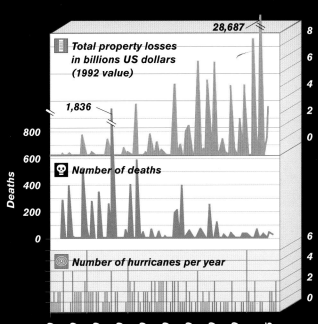

28,687

Total property losses in billions US dollars (1992 value)

1,836

Property losses: 8 · 6 · 4 · 2 · 0

Deaths: 800 · 600 · 400 · 200 · 0

Number of deaths

Number of hurricanes per year

Hurricanes: 6 · 4 · 2 · 0

1900 · 1910 · 1920 · 1930 · 1940 · 1950 · 1960 · 1970 · 1980 · 1995

ENSO

The far-reaching consequences of changes in the Pacific

The phrase El Niño ("Christ Child," in Spanish) was coined in the 19th century by Peruvian fishermen to describe the warm sea current that occasionally appeared at Christmas. Now known as the El Niño-Southern Oscillation (ENSO), it is one of the most important elements in the Earth's climate, its effect being felt far and wide across the globe; but are we any closer to being able to understand it, let alone predict when it will occur?

18 June, 1999

12 November, 1999

8 January, 2000

11 March, 2000

PREDICTING EL NIÑO

There is a large network of ocean and satellite monitoring systems over the Pacific Ocean, aimed primarily at recording sea-surface temperature, the major factor in the El Niño-Southern Oscillation (ENSO). By constructing both computer circulation and statistical models based on this climatic data, we are steadily getting better at predicting the likelihood of an El Niño or a La Niña event. But although we've made great progress, our understanding of the ENSO phenomenon is really still in its infancy—as, too, is our ability to predict when one will occur.

Left *These satellite pictures show the evolution of an El Niño event from June 1999 to March 2000. Note the warm water spreading slowly across the Pacific Ocean, which is typical of an El Niño event.*

EL NIÑO IN 1997–1998

The 1997–1998 El Niño conditions were the strongest on record and caused droughts in the southern United States, East Africa, northern India, north-east Brazil, and Australia. In Indonesia, forest fires burned out of control in the very dry conditions. In California, parts of South America, Sri Lanka, and east central Africa, there were torrential rains and terrible floods.

Below *Many areas suffer severe droughts during El Niño events.*

CORAL REEFS AND EL NIÑO

Coral reefs are a valuable economic resource for fisheries, recreation, tourism, and coastal protection, and some people estimate that the global cost of losing the coral reefs runs into hundreds of billions of dollars each year. In addition, reefs are one of the largest global stores of marine biodiversity, with untapped genetic resources. Tragically, coral reefs are directly threatened by global warming.

The last few years have seen an unprecedented decline in the health of coral reefs. In 1998 El Niño was associated with record sea-surface temperatures and associated coral bleaching. This involves the coral expelling the algae that live within them and are necessary for their survival. In some regions as much as 70 percent of the coral may have died in a single season. There has also been an upsurge in the variety and virulence of coral diseases in recent years, with major die-offs in Florida and much of the Caribbean region. In addition, increasing atmospheric carbon dioxide concentrations could decrease the calcification rates of the reef-building corals, resulting in weaker skeletons, reduced growth rates, and increased vulnerability to erosion. Model results suggest these effects would be most severe at the current margins of coral-reef distribution. Global warming is bad news for coral reefs.

Above *The hot surface waters during an El Niño event can bleach or kill corals. Here dead corals off the Maldives lie littered on the sea floor.*

ENSO AND GLOBAL WARMING

Already, there are fears that ENSO has been affected by global warming. El Niño conditions should only occur every three to seven years, but between 1991 and 1995 they returned for three years out of four. Then they occurred again in 1997–98.

Scientists at the University of Colorado have discovered evidence in coral reefs of two major changes in the frequency and intensity of El Niño events. Coral calcifies and builds up over a period of many years, so, by drilling a core in the coral reefs in the Western Pacific, the scientists were able to assess the state of the coral over a period of some 150 years–well beyond our historical records–in much the same way as one can gauge information about climatic conditions by looking at the growth rings in a tree's trunk.

The coral provided evidence of changes in sea-surface temperature. These changes indicate shifts in ocean current, which accompanies shifts in the ENSO. First, at the beginning of the 20th century, there was a shift from a cycle lasting 10 to 15 years, to one lasting 3 to 5 years; then, in 1976, there was a marked shift to more intense and even more frequent El Niño events.

Considering the huge weather disruption and disasters caused by El Niño, these are sobering facts. Computer models also suggest that the current "heightened" state of El Niño can permanently shift weather patterns: it seems, for example, that the drought region in the United States could well be shifting eastward.

However, as we have seen, it is hard enough to predict whether there will be an El Niño event in six months' time, never mind whether or not ENSO is going to become more extreme over the next 100 years. We have a long way to go.

Above *Corals grow so slowly that by drilling them we can reconstruct what the ocean was like tens and even hundreds of years ago. This gives us a clearer picture of how and when the frequency of El Niño has changed.*

MONSOONS: SOUTH AMERICA

Amazon monsoon brings torrential rain

The Amazon River—so immense it discharges one-fifth of all freshwater running into the oceans of the world—is a product of the monsoon. The Amazon drainage basin is the largest on Earth, covering an area about the size of Europe. The rainforest is spectacular and supports a huge number and diversity of species.

In 1542 Francisco de Orellana led the first European voyage down the Amazon River. During this intrepid voyage the expedition met a lot of resistance from the local Indians; in one particular tribe, the women warriors were so fierce that they drove their male warriors in front of them with spears. Thus the river was named after the famous women warriors of the Greek myths, the Amazons. It was this voyage that started our almost mystical wonder at the greatest river and the largest area of rainforest in the world.

The Amazonian rainforest is a direct product of the monsoons. Every year they bring 6.5 feet (more than 2m) of rain. The rivers—more than 1,000 of them in addition to the mighty Amazon itself—overflow and inundate vast areas, bringing nutrient-rich sediments that sustain the flora and fauna of the most diverse natural habitat on Earth.

The rainforest is particularly important when it comes to the future of global warming, as it is a

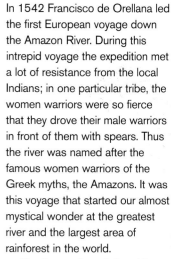

Left *The world's rainforests are hot spots of biodiversity—essential if we are to maintain the genetic heritage of the Earth.*

huge natural store of carbon. Until recently, it was thought that an established rainforest such as the Amazon had reached maturity and thus could not take up any more carbon dioxide. Experiments have shown this to be wrong: the Amazon rainforest is sucking up five tons of atmospheric carbon dioxide every 2.4 acres (1 hectare) per year. This is because plants react favorably to increased carbon dioxide, which is the raw material for photosynthesis, so having more carbon dioxide in the atmosphere acts like a fertilizer, stimulating plant growth. It seems that the Amazon rainforest is presently absorbing a large percentage of our atmospheric carbon dioxide pollution—equivalent to about three-quarters of the world's car pollution. But things could change.

CHANGE AND DESTRUCTION

Climate models developed at the UK Meteorological Office's Hadley Centre for Climate Prediction and Change suggest that, by 2050, global warming could have increased the winter dry season. According to these models, global warming will cause the climate to shift toward a more El Niño-like state, with a much longer South American dry season. This would destroy the Amazon rainforest, which requires not only a large

Right Scientists fear that global warming could turn much of the Amazon rainforest into savanna which is found both to the east south east of Amazonia and in Patagonia.

amount of rain during the wet season, but a relatively short dry season. Extended dry periods would lead to fire demolishing large parts of the rainforest; it would be replaced by savanna (dry grassland), which is adapted to coping with the long dry season. This savanna is already found both to the east and south of the Amazon Basin.

Savanna has a much lower carbon storage potential than rainforest, thus the carbon stored in the rainforest would be sent back into the atmosphere, accelerating global warming. The rainforest currently helps to reduce the amount of pollution we put into the atmosphere, but may ultimately cause global warming to accelerate at an unprecedented rate.

Below The Amazon River, by far the largest river in the world. The river and the rainforest that it supports are interdependant: if global warming reduces the monsoon rains, the forest, too, will suffer.

MONSOONS: BANGLADESH & EGYPT

Deltas at risk

If we continue our tradition of environmental abuse, the global sea level could rise between 8 and 35 inches (20 and 88 cm) in the next 100 years, primarily due to the thermal expansion of the oceans. This will have serious ramifications worldwide, particularly in highly populated, low-lying delta areas such as Bangladesh and Egypt.

Right *The mouth of the River Ganges seen from space clearly shows what much of Bangladesh is like, with networks of ever smaller and smaller rivers branching off from each other. The combination of silt and fresh water brought in by the Ganges makes this area one of the most fertile in the world.*

BANGLADESH: POTENTIAL IMPACT OF SEA-LEVEL RISE

PRESENT SEA LEVEL
Popluation: 112 million
Land area: 51,724 square miles
(134,000 km²)

Dhaka
(Dacca)

BANGLADESH

Bay of Bengal

Above *As a result of the combined effects of sea-level rise and reduced sediment flowing into the delta from the country's major rivers, Bangladesh could lose up to 16 percent of its land.*

POTENTIAL SEA-LEVEL RISE OF 5 FEET (1.5 m)
Popluation affected: 17 million (15%)
Land area affected: 8,492 square miles
(21,990 km²)(16%)

Rising sea levels are a major concern in all coastal areas, as this will decrease the effectiveness of coastal defenses against storms and floods and increase the instability of cliffs and beaches. In the United States and the United Kingdom, the response to this danger has been to raise the height of sea walls around property on the coast; to abandon some poorer quality agricultural land to the sea (as it was no longer worth the expense of protecting it); and to add further legal protection to coastal wetlands which are nature's best defense against the sea.

Elsewhere in the world, however, there are countries which face much more serious threats from rising sea levels. For small island nations such as the Maldives in the Indian Ocean and the Marshall Islands in the Pacific, a 3-foot (1-m) rise in sea level

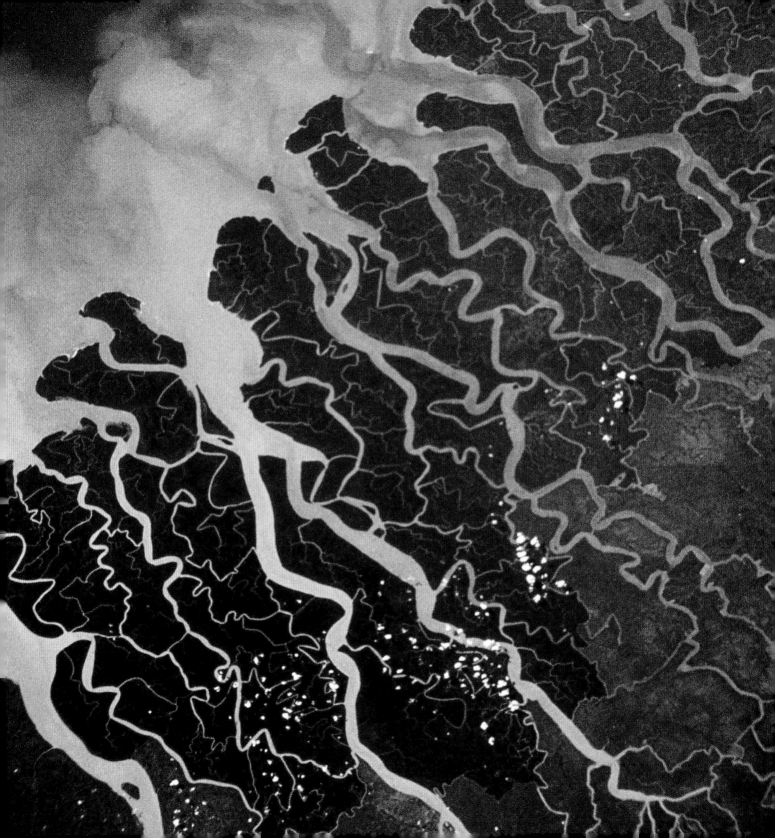

would flood up to 75 percent of the dry land, making the islands uninhabitable.

There is a further twist to the story if we consider nations with a significant portion of their country on river deltas—for example, Bangladesh, Egypt, Nigeria, and Thailand. A World Bank report in 1994 concluded that other human activities on the deltas were exacerbating the effects of global warming by causing these delta areas to sink, increasing their vulnerability to storms and floods.

BANGLADESH

In the case of Bangladesh, over three-quarters of the country is within the deltaic region formed by the confluence of the Ganges, Brahmaputra, and Meghna rivers. This fertile Bengal Delta supports one of the world's most dense populations, over 110 million people in 54,000 square miles (140,000 square kilometers). Over half the country lies less than 17 feet (5 m) above sea level, and thus flooding is a common occurrence. During the summer monsoon a quarter of the country is flooded. These floods, like those of the Nile, bring with them life as well as destruction: the water irrigates and the silt fertilizes the land. But the monsoon floods have been getting worse throughout the 1990s.

Every year the Bengal Delta should receive over one billion tons of sediment and 385 cubic miles (1,000 cubic kilometers) of freshwater. This sediment load balances the erosion of the delta

Above The aftermath of the flooding in Bangladesh in 1998 meant that people were not able to return to what was left of their homes for many months, until the flood waters finally receded.

Right The Ganges is a source of international controversy as the Indian government has dammed the river to provide water for irrigation. This reduces the amount of silt and water flowing into Bangladesh and is starting to erode the land.

Below right Alexandria, one of Egypt's major ports, is, like much of northern Egypt, under threat from rising sea levels caused by global warming.

NILE DELTA:
POTENTIAL IMPACT OF SEA-LEVEL RISE

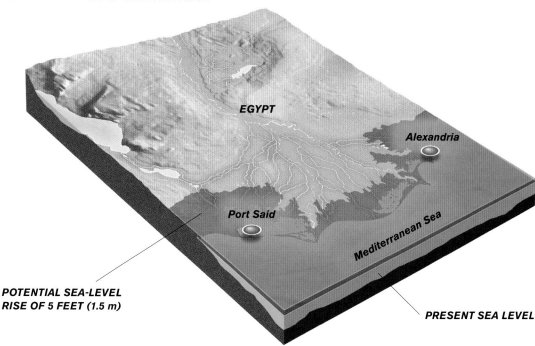

EGYPT

Alexandria

Port Said

Mediterranean Sea

**POTENTIAL SEA-LEVEL
RISE OF 5 FEET (1.5 m)**

PRESENT SEA LEVEL

less than 7 feet (2 m) above sea level and is only protected from flooding by a coastal sand belt 0.5–6 miles (1–10 km) wide, shaped by the Rosetta and Damietta branches of the Nile. Erosion of the sand belt has accelerated since the momentous building of the Aswan Dam in the south of Eygpt.

Rising sea levels would destroy weak parts of the sand belt, which are essential for the protection of lagoons and the low-lying reclaimed lands. The impacts would be far reaching. About one-third of Egypt's fish catches are made in the lagoons, and a rise in sea level would change the water quality and affect most freshwater fish. Valuable agricultural land would be inundated. Vital, low-lying installations in Alexandria and Port Said would be threatened. Tourism in seaside areas would be endangered and essential groundwater would be salinated.

Dykes and protective measurements would probably prevent the worst flooding as long as the rise in sea level did not amount to more than 20 inches (50 cm). Nonetheless, the impact of groundwater salination and increased wave action on coastal areas would be substantial.

by both natural processes and human activity. In neighboring India, however, the Ganges River has been diverted into the Hooghly Channel for irrigation purposes. As a result, less sediment is washed down into Bangladesh during the monsoon, causing the delta to subside. The extraction of freshwater from the delta for agricultural use and for drinking water is exacerbating the situation.

In the 1980s, 100,000 tubewells and 20,000 deep wells were sunk, increasing the freshwater extraction sixfold. These projects, which were aimed at increasing the quality of life for people in the region, have produced a subsidence rate of up to 1 inch (2.5 cm) per year, one of the highest in the world. In the worst-case scenario, using estimates of subsidence rate and sea-level rise due to global warming, the World Bank has estimated that by the middle of the 21st century the sea level in Bangladesh could rise by as much as 6 feet (1.8 m). This would result in a loss of up to 16 percent of land, which currently supports 13 percent of the population and produces 12 percent of the current gross domestic product (GDP). This does not take into account the devastation of the mangrove forests and the associated fisheries. Moreover, saltwater flowing into inland areas would further damage water quality and agriculture.

Although Bangladesh is the worst case globally, similar changes can be observed at all other major delta regions.

THE NILE DELTA

Another example of a threatened coastline is the Nile Delta, which is one of the oldest intensely cultivated areas on Earth. It is very heavily populated, with population densities of up to 4,800 inhabitants per square mile (1,600 per square kilometer).

Only the Nile Delta and the Nile Valley, which represent 2.5 percent of Egypt's land area, are suitable for intensive agriculture; deserts surround the low-lying, fertile flood plains. Most of a 30-mile (50-km) wide landstrip along the coast is

DEEP WATER FAILURE

Ocean circulation is a major climate control factor

In fact, the deep ocean is the only candidate for driving and sustaining long-term climate change (over hundreds or thousands of years) because of its volume, heat capacity, and inertia.

Two deep ocean currents—one originating in the North Atlantic Ocean and the other in the Southern Ocean—play a major role in the Earth's climate. As always in nature, the balance between the two is critical.

NORTH ATLANTIC DEEP WATER (NADW)

In the North Atlantic, the Gulf Stream carries warm and salty surface water northward from the Gulf of Mexico to the Nordic seas. The increased saltiness or salinity of the Gulf Stream is due to high levels of evaporation in the Caribbean, which removes moisture from the surface waters and concentrates the salts. As the Gulf Stream flows northward, it cools down. The combination of a high salt content and a low temperature makes the surface water heavier, or denser. By the time it reaches the relatively fresh oceans north of Iceland, it has become dense enough to sink into the deep ocean. The "pull" exerted by this dense, sinking water helps to maintain the strength of the Gulf Stream, ensuring a flow of warm tropical water into the North Atlantic that sends mild air masses across to Europe.

Above *The ocean acts like a giant conveyor belt: warming water flowing northward in the Atlantic Ocean sinks down and flows all the way to the Antarctic Ocean and then around into the other oceans. There it rises and flows back to the North Atlantic as surface currents.*

DEEP OCEAN CURRENTS IN THE ATLANTIC

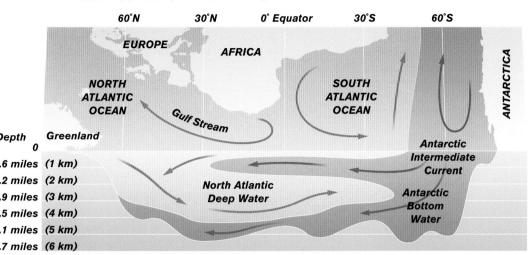

Above *The deep ocean circulation is a fine balance between the North Atlantic Deep Water and the colder, denser Antarctic Bottom Water. We now know that the northern and southern deep waters are inextricably linked, and variations in one directly influence the other.*

The newly formed deep water sinks to between 6,500 and 11,500 feet (2,000 and 3,500 m) below the surface of the ocean and flows southward down the Atlantic Ocean. This current is known as the North Atlantic Deep Water (NADW).

ANTARCTIC BOTTOM WATER (AABW)

In the South Atlantic Ocean the NADW meets a second type of deep water which is formed in the Southern Ocean and is called the Antarctic Bottom Water (AABW). Antarctica is surrounded by sea

ice. There are large holes in this sea ice, produced by out-blowing Antarctic winds which push the sea ice away from the continental edge, and deep water forms in them. The winds are so cold that they supercool the exposed surface waters. This leads to the formation of more sea ice and salt rejection (no one has yet been able to produce a salty ice cube!), producing the coldest and saltiest water in the world. The AABW flows around Antarctica and penetrates the North Atlantic, flowing under the warmer and thus somewhat lighter NADW. The AABW also flows into both the Indian and the Pacific Oceans.

MAINTAINING THE BALANCE

The balance between the NADW and the AABW is extremely important in maintaining our

THE GULF STREAM

It has been calculated that the Gulf Stream delivers 27,000 times the energy of all the UK's power stations put together. If you are in any doubt about how good the Gulf Stream is for the European climate, compare the winters at the same latitude on the two sides of the Atlantic Ocean—for example, London with Labrador, or Lisbon with New York.

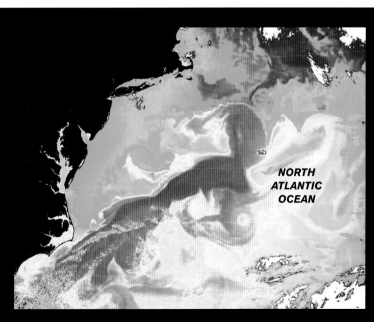

Left *The warm Gulf Stream is a massive surface water current of 200,000,000 gallons (909,200,000 liters) of water flowing per second. The Gulf Stream ensures that warm tropical water flows northward, sending mild air masses across the European continent.*

Left *Key to predicting our future climate is knowing whether the Antarctic (pictured left) or the Arctic will melt first.*

Below *Successive satellite pictures showing two massive icebergs, C-16 and B-15A, breaking off over the Antarctic summer of 2000–2001.*

weather. However, the influence of the Gulf Stream is mainly in the winter so it does not affect summer temperatures. So if the Gulf Stream fails summer temperatures will not be affected and global warming would still cause these to rise. Thus Europe would end up with extreme seasonal weather.

Conversely, if the Antarctic ice sheet starts to melt significantly before the Greenland and Arctic ice, things could be very different. If enough meltwater is put into the Southern Ocean, then the AABW will be severely curtailed. Because

present climate, as not only does it keep the Gulf Stream flowing past Europe, but it also maintains the right amount of heat exchange between the Northern and the Southern Hemispheres.

Scientists have shown that the circulation of deep water can be weakened or "switched off" if there is enough input of fresh water to make the surface water too light to sink. They have coined the phrase "dedensification" to describe this phenomenon.

As we have seen, there is already evidence that global warming is causing significant melting of the polar ice caps in both the Arctic and Antarctica. This will lead to a lot of freshwater being added to the polar oceans. Global warming could therefore cause the collapse of the NADW, weakening the warm Gulf Stream. This would cause much colder European winters, stormier conditions, and more severe

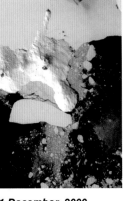

11 December, 2000

29 December, 2000

5 January, 2001

16 January, 2001

15 February, 2001

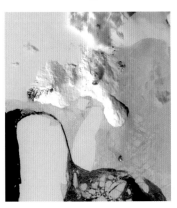

9 December, 2001

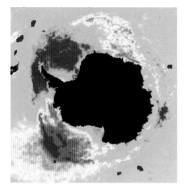

Above *The amount of sea ice around Antarctica is the key measure of whether the continent is significantly warming up due to global warming. Satellites now record the extent of sea ice around Antarctic all year round to monitor any major changes.*

BRITISH ISLES

KEY

INCREASE IN

Flooding

Coastal erosion

Heavy rain

Storms

Gales

Exotic marine fauna

Mediterranean climate

Holidays and sunbathing

Forest fires

Tropical diseases

Left *The effect on the UK if the Arctic were to melt significantly before the Antarctic, preventing the formation of the North Atlantic Deep Water and consequently affecting the Gulf Stream.*

Below *The effect on global sea level if AABW production were curtailed. The NADW would spread out throughout the world's oceans to replace it and, because this northern deep water is about 9°F (5°C) warmer than its southern counterpart, it would take up more space, causing a staggering 6.5–10-foot (2–3-m) rise in sea level with catastrophic consequences around the world.*

the deep water system is a delicate balancing act between NADW and AABW, if the AABW is reduced then the NADW will increase and expand, as the NADW is warmer than the AABW (and if you heat up a liquid it expands). So any increase in the NADW will mean a massive increase in sea level, of up to 8 feet (2.5 m).

WILL IT HAPPEN?

The problem is that we have no idea how much freshwater it will take to shut off either the NADW or the AABW, nor can we predict whether the Arctic or Antarctica will melt first. What we do know is that partial melting has happened in the past and, if global warming continues, sometime in the future there will be either a severe alteration in the European climate or a significant rise in sea levels.

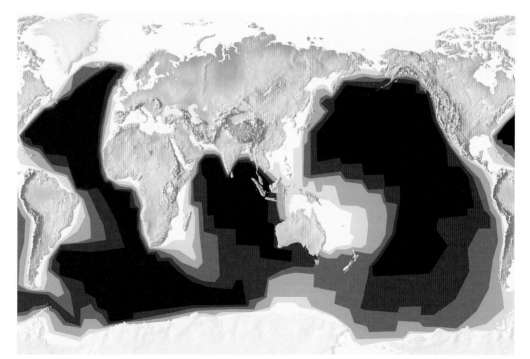

KEY

■ 200 inches (500 cm)	■ 80 inches (200 cm)	40 inches (100 cm)
■ 100 inches (250 cm)	60 inches (150 cm)	20 inches (50 cm)

GAS HYDRATES

The deadly threat

Beneath the world's oceans and permafrost lurks the menace of gas hydrates. They are extremely unstable and, if released into the atmosphere, could significantly affect global warming.

Gas hydrates are a mixture of water and methane sustained as a solid at very low temperatures and very high pressures. They are composed of a cage of water molecules, frozen as ice beneath the deep oceans and the permafrost, which hold individual molecules of methane. The methane comes from decaying organic matter found deep in ocean sediments and in soils beneath permafrost. These gas hydrate reservoirs are extremely unstable–a slight increase in temperature or decrease in pressure can cause them to destabilize. The reservoirs therefore pose a major risk as global warming will heat up both the oceans and the permafrost and could cause the ice surrounding the methane to melt, pumping huge amounts of methane into the atmosphere. Methane is 21 times more powerful than carbon dioxide. If enough is released it would raise temperatures even more, releasing even more gas hydrates–a runaway greenhouse effect.

HISTORICAL EVIDENCE

There are 10,000 giga tons of gas hydrates stored beneath our feet, compared with only 180 giga tons of carbon dioxide currently in the atmosphere. Worryingly, there is evidence that a runaway greenhouse effect occurred 55 million years ago. During this hothouse event only 1,200 giga tons of gas hydrates were released–but that was enough to accelerate the natural greenhouse effect, producing an extra 9°F (5°C) of warming.

WORLD DISTRIBUTION OF GAS HYDRATES

KEY

- Seafloor deposits
- Permafrost deposits

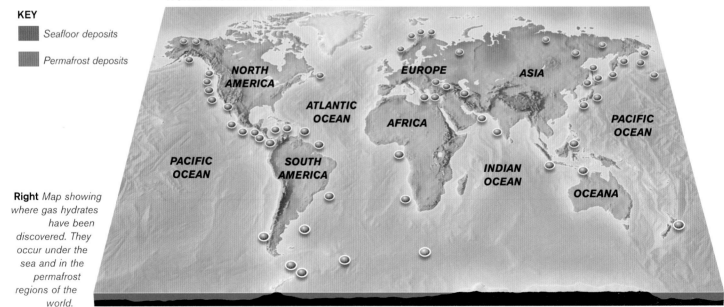

Right *Map showing where gas hydrates have been discovered. They occur under the sea and in the permafrost regions of the world.*

THE SITUATION TODAY

One hope is that the current global warming will not warm the deep ocean sufficiently to cause the release of gas hydrates. There is, however, a second possibility. If a significant proportion of Greenland and/or Antarctica melts, a huge weight will be removed from the land as the ice disappears. If this happens, the land that has been constantly pushed down by the huge weight of ice will start to move upward relative to the surrounding ocean and land.

This phenomenon of land bouncing back up after the ice weight has been removed is called "isostatic rebound." If it happens to

Greenland and/or Antarctica, it will mean the relative sea level around the continental shelf will become lower, removing the weight of the water and thus pressure on the deep-sea sediment. Pressure removal is an extremely efficient

way of destabilizing gas hydrates and hence huge amounts of methane could be released into the atmosphere from around the Arctic and Antarctica.

THE BRITISH ISLES— A CLASSIC EXAMPLE

During the last ice age, Scotland was covered by a huge ice sheet over 1-mile (1.6-km) deep. This exerted immense pressure and pushed down the land. As Scotland and England are obviously connected, pushing down on Scotland is like pushing down on a see-saw—and so England was swung upward by Scotland's depression. Since the removal of the ice sheet from Scotland 10,000 years ago, the land has slowly recovered and is constantly moving very slowly upward. Because of the see-saw effect, England is correspondingly moving downward.

Above *Scotland is still recovering from the effects of the last ice age. Pushed down by the huge ice sheet, it is only slowly returning to its original height.*

GAS HYDRATE-GENERATED TSUNAMIS

A secondary effect of gas-hydrate release is that when it breaks down, it can do so explosively. There is clear evidence that gas-hydrate release in the past has caused massive slumping of the continental shelf and massive tsunamis (giant sea waves). The most famous is the Norwegian Storegga slide, which occurred about 8,000 years ago. It produced a 50-foot (15-m) tsunami that wiped out many Scottish coastal settlements.

We cannot rule out the possibility that global warming could lead to an increased frequency of massive killer waves of 50 feet (15 m) in height hitting our coasts. Up to now, only the countries around the Pacific Rim are prepared for this type of event as many of these tsunamis are set off by earthquakes. But gas hydrate-generated tsunamis could occur anywhere in the ocean.

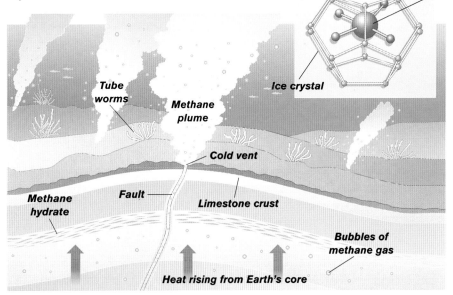

Above *How gas hydrates release methane into the atmosphere. The methane molecule (see inset) is trapped in a cage of ice; once this melts, the methane is released and makes it ways up into the* atmosphere via faults in the Earth's crust. If too much methane is released, it could increase global warming. This, in turn, would release more trapped methane, increasing global warming still further.

WHAT CAN WE DO ABOUT WARMING?

Cut emissions!

The most logical approach to the global warming problem would be to significantly cut emissions. Unfortunately, this has major implications for the world economy.

THE KYOTO PROTOCOL

The United Nations Framework Convention on Climate Change (UNFCCC) was created at the Rio Earth Summit in 1992 to try and negotiate a worldwide agreement in reducing greenhouse gases and limiting the impact of global warming.

Since then, two major steps forward have been achieved. The first occurred at midnight on December 13, 1997 when the Kyoto Protocol was formed, which set out the general principles for a worldwide treaty on cutting greenhouse gas emissions.

The Kyoto Protocol stated that all developed nations would aim to cut their emissions by 5.2 percent of their 1990 levels by the years 2008–2012. However, some countries have continued to increase their carbon dioxide emission significantly since 1990: the United States, for example, now produces 30 percent more carbon dioxide pollution than it did in 1990, so in order to agree with and ratify the Kyoto Protocol, it would have to cut its emission by over a third, which would be damaging to its economy.

The second breakthrough occurred in Bonn on July 23, 2001 when 186 countries ratified and signed the Kyoto Protocol, making it a legal treaty. However, the United States, under the leadership of President Bush, had withdrawn from the climate negotiations in March 2001 and thus did not sign the Kyoto Protocol at the Bonn meeting. With the United States producing about a quarter of the world's carbon dioxide pollution, this was a big blow. Moreover the targets set by the Kyoto Protocol were reduced during the Bonn meeting to make sure that Japan, Canada,

Below *More and more people are realizing that something must be done about global warming. Many people are shocked by President Bush's decision to withdraw the United States from the Kyoto Protocol.*

SO WHAT HAVE THE 186 NATIONS SIGNED UP FOR?

■ The 38 industrialized nations have agreed to binding targets to reduce their greenhouse gas emissions.

■ The EU will immediately start turning the treaty into law for all member countries, forcing an overall 8 percent cut in greenhouse gas emissions by the year 2010. The United Kingdom's target will be 12.5 percent, thus allowing poorer EU countries a lower target.

■ US$500 million of new funds a year will be provided by the industrialized world to help developing countries adapt to climate change and to provide new, clean technologies.

■ Industrial countries will be able to plant forests, manage existing ones, and change farming practices, and thereby claim credit for removing carbon dioxide from the atmosphere.

■ An international trade in carbon will be started. Companies saving carbon by building clean technologies in other states will be able to claim credits which can be sold as tons of carbon saved on an international commodity market. This will most likely be set up in London.

■ If countries fail to reach the first set of targets by 2012, they will have to add the shortfall, plus an extra penalty of 30 percent of however much they failed to cut, to the next commitment period . They will also be excluded from carbon trading and be forced to take corrective measures at home.

Left *We face a stark future with industrial pollution making droughts and floods more common around the world.*

and Australia would join. The targets for the 37 richest and most developed countries will be a cut of about 1 to 3 percent compared with their 1990 levels.

However, the treaty does not include underdeveloped countries. This is worrying because, if countries such as India and China continue to develop, they will produce huge amounts of pollution. If these two countries achieve the same car-to-family ratio as Europe then there will be an extra billion cars in the world!

The Kyoto Protocol agreed at Bonn could "enter into force" as early as 2002, once at least 55 countries representing more than 55 percent of the global emissions have passed it as national law.

FINANCIAL IMPLICATIONS

We have discussed the global warming impacts on the environment, but it is important to realize that climate change could have huge financial implications.

Currently natural disasters cost insurance companies US$10 billion per year, with the real economic losses being more than US$100 billion. Predicted financial losses for the future are even bigger and will probably be measured in percentage points of the world's gross product.

The irony pointed out by some researchers is that the major financial players who will bear the brunt of these losses (insurers, banks, and pension holders), are, by virtue of share holdings, the ones who own the oil companies that are exacerbating global warming and causing the financial losses. Somehow the link between fossil fuel-related carbon dioxide emissions and the financial sector must be broken. Investment in

Above *In 1992 at the Rio Earth Summit, the United Nations Framework Convention of Climate Change (INFCCC) was set up. Since then, this organization has collated and reported all the relevant science on global warming.*

renewable and non-carbon dioxide energy sources has to start—and it has to start on a global scale. Strangely enough, the insurance companies may be the ones who have enough financial and political weight to start the whole process.

INSURANCE IN A STORMIER WORLD?

One of the most worrying things about global warming is the greater unpredictability of the weather and enhanced strength of an increasing number of storms. It has been shown that natural catastrophes worldwide have risen threefold since the 1960s.

It is in the interest of the insurance industry to try and predict the chance of disasters happening. The year 1998—officially the warmest on record—cost the industry US$93 billion in damages. A third of this was from the storms and flooding in China, which killed 3,700 people. This was the second highest ever loss for the insurance industry, topped only in 1995, when an earthquake in Kobe, Japan, took costs spiraling to US$180 billion.

Payouts by insurers for damage caused by natural disasters are now 15 times higher than they were 25 years ago. As we go into the new millennium, we are entering an era when insurers can neither predict nor cover all the damage caused by the effects of global warming. In some cases, in areas known to be most at risk, home owners will no longer be able to get insurance or, if it exists, it will be unaffordable.

The following catalog will show you why:
- The El Niño event of 1998 is said to have caused 80 separate catastrophes, which resulted in US$14 billion of damage of which only US$2 billion was insured.
- During the hailstorms that swept the American Midwest in May 1998, only one person died but there was US$1.8 billion of damage.
- Insurance companies are worried about the huge losses in 1999 caused by the massive storms in Europe: 2,800 billion cubic feet (80 billion cubic meters) of snow fell on the Alps, enough to raise the level of Lake Constance by 40 feet (12 m), which caused huge floods.
- In 2000 and 2001, insurance companies were hit by huge claims after the devastating floods in Italy and Britain.

Right *Global warming could increase the number and strength of storms, providing an increased risk to both agriculture (top) and coastal communities (below).*

TECHNOFIXES

Can we fix global warming?

Governments worldwide have shown a lot of interest in "alternative" or "technofix" solutions to the problem of global warming.

We have seen that governments are slowly getting their act together and beginning to seriously investigate ways of reducing carbon dioxide emissions.

There are four main areas of technofixes:

■ Carbon dioxide removal from industrial processes can contribute substantially to a reduction in atmospheric carbon dioxide, but further research and development is required to improve methods and ensure that they can be applied within the concepts of sustainable development.

■ We can use less energy and thus produce less carbon dioxide. It is feasible to improve energy efficiency by 50 percent on average over the next three decades. The conversion efficiency in power generation, for example, can be increased by 60 percent using advanced technologies in the field of gas turbines and fuel cells. However, this would require tough policy measures such as the introduction of a high energy or carbon tax.

■ We should adopt renewable energy sources—that is, energy sources that do not produce carbon dioxide. Most promising in the short term is biomass, which by the year 2020 could produce one-third of the world's energy requirements. Most promising in the long term, however, is solar energy.

■ We can aid the removal of carbon dioxide from the atmosphere either by growing new forests or by stimulating the ocean to take up more (see Iron Hypothesis, opposite).

Below left *One way to prevent global warming is to reduce the amount of carbon dioxide emitted by industrial processes.*

Below right *Planting more forests would take more carbon dioxide out of the atmosphere, reducing the effect of global warming.*

HAZARDOUS SOLUTIONS

All of the aforementioned methods make sense and a combination could be used to combat global warming. At the same time, however, they all have their drawbacks.

Removing carbon dioxide during industrial processes is tricky and costly, because not only does the carbon dioxide need to be removed but much of it must be stored somewhere as well. Removal and storage costs could be somewhere between US$20 and US$50 per ton of carbon dioxide. This would cause a 35 to 100 percent increase in power production costs. (It is worth noting in passing that uses can be found for some recovered carbon dioxide: some may be utilized in enhanced oil recovery, the food industry, chemical manufacturing [producing soda ash, urea, and methanol], and the metal processing industries. Carbon dioxide can also be used in the production of construction materials, solvents, cleaning compounds, and packaging, and in wastewater treatment. But in reality, most of the carbon dioxide captured from industrial processes would have to be stored.)

Theoretically, two-thirds of the carbon dioxide formed by the combustion of the world's total oil and gas reserves could be stored in the corresponding reservoirs. Other estimates indicate storage of 90–400 giga tons of carbon (GtC) in natural gas fields alone and another 90 GtC in aquifers. Oceans could also be used to dispose of the carbon dioxide. Suggestions have included storing

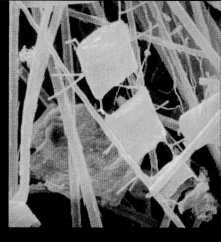

IRON HYPOTHESIS

Global warming is constantly on the political agenda, even if politicians do not like to mention it. The problem, however, is that cutting carbon dioxide emissions has a huge economic price tag—so scientists and politicians are always looking for a quick fix or a "technofix" for global warming.

The late Professor John Martin of Moss Landing Marine Laboratories in California put forward one of the most controversial ideas. He has suggested that much of the

Above *Phytoplankton are tiny plants—the basis for the entire marine food chain. Their growth is becoming limited in oceans which lack iron in the surface water.*

world's oceans are underproducing because of a lack of the vital nutrients that allow plants to grow in the surface waters. The most important of these nutrients is iron. Marine plants need minute quantities of iron in order to grow, and in most oceans enough iron-rich dust falls from the land. However, it seems that large areas of the Pacific and Southern Pacific Ocean do not receive much dust and thus are barren of iron—and so it has been suggested that we could fertilize the ocean with iron to stimulate marine productivity.

The extra photosynthesis would convert more surface-water carbon dioxide into organic matter. When the organisms die, the organic matter would drop to the bottom of the ocean, taking with it the extra carbon which would, effectively, be stored there. The reduced surface water carbon dioxide would be replenished by carbon dioxide from the atmosphere. So, in short, fertilizing the world's oceans would help to remove atmospheric carbon dioxide and store it in deep-sea sediments. However, the amount of iron required is huge and, as soon as you stopped adding the extra iron, a lot of the stored carbon dioxide would be released.

There is also another, darker side to this iron hypothesis. It seems that, because of industrialization and worldwide land-use changes, there is about 150 percent more dust in the atmosphere than there was 200 years ago. This extra dust has increased the oceans' ability to take carbon dioxide out of the atmosphere—so our dirty atmosphere is literally helping us against global warming. However, under the Kyoto Protocol, countries are going to start expanding forests and preventing soil erosion to draw carbon dioxide out of the atmosphere. This will ultimately lead to a decrease in dust. Calculations by Dr. Andrew Ridgwell at the University of East Anglia and myself suggest that a significant proportion of the extra carbon dioxide stored on land could, under the Kyoto Protocol, be returned to the atmosphere because of

SCRIPPS CARBON DIOXIDE

SCRIPPS IR ANALYZER

AN EXPLOSIVE OVERTURN

An example of the dangers of storing carbon dioxide occurred in 1986, when a tremendous explosion of carbon dioxide from Lake Nyos, in the north-west of Cameroon, killed more than 1,700 people and livestock up to 15 miles (25 km) away. Though similar disasters had occurred previously, never before had one heard of Mother Nature asphyxiating human beings and animals on such a scale in a single brief event.

What happened was this: Lake Nyos lies in the mouth of a dormant volcano. For years, dissolved carbon dioxide from the volcano had been seeping from springs beneath the lake and was trapped in deep water by the weight of water above. In 1986, an avalanche sent tons of rocks crashing into the lake which churned up the lake's waters, resulting in the trapped carbon dioxide rising to the lake's surface and escaping into the atmosphere—with fatal consequences.

carbon dioxide by hydrate dumping—that is, by mixing carbon dioxide and water at high pressure and low temperatures to create a solid or hydrate that is heavier than the surrounding water in the ocean, and thus drops to the bottom. This hydrate is very similar to the methane hydrates discussed on pages 118 and 119.

The major problem with all these storage methods is safety. Carbon dioxide is a very dangerous gas—it is heavier than air and causes suffocation. Hence storing carbon dioxide is very difficult and potentially lethal. With ocean storage there is the added complication that the oceans circulate—so whatever carbon dioxide you dump in there will eventually return. Moreover, scientists are very uncertain about

the environmental effects on the oceans' ecosystems. Therefore there are no estimates of the amount of carbon dioxide that can be safely stored.

From the safety and environmental perspective the storage of carbon dioxide either underground and/or in the ocean is really not feasible, however helpful this would be in the short term. Ultimately a combination of improved energy efficiency and alternative energy sources are the solutions to global warming.

Top Greenhouse gas research. Scientist notes the level of carbon dioxide in air from a chart record produced by the Scripps infrared gas analyzer. This has shown that carbon dioxide levels are rising at just over one part per million each year due to the burning of fossil fuels.

Right *Cameroon is situated in West Africa. The region is volcanically active.*

Below left *The region around Lake Nyos depended largely on agriculture: both people and livestock were wiped out in the tragedy.*

Below *Lake Nyos lies in the mouth of a dormant volcano. In 1986, carbon dioxide released from the lake killed over 1,700 people. The picture below illustrates the huge amount of mud churned up by this explosion.*

NIGERIA

CAMEROON

● Lake Nyos

Bight of Biafra

●Yaoundé

ATLANTIC OCEAN

ADAPTATION AND MITIGATION

Action reduces impact

If we can predict what the impacts of global warming are likely to be, national governments can take action to mitigate the effects by, for example, introducing new laws.

Below *We face a stark future, with industrial pollution making droughts and floods more common around the world*

Worldwide, carbon dioxide emissions must be cut by between 60 and 80 percent. As the Kyoto Protocol will only result in a cut of between 1 and 3 percent, the second major aim of the Intergovernmental Panel on Climate Change (IPCC) is to study and report on the sensitivity, adaptability, and vulnerability of each national environment and its socio-economic system.

Above *The Earth Summit in Rio in 1992. It was recognized that any solution to global warming would have to consider the needs of the poorest nations: their right to industrial development and the need to move toward a sustainable economy.*

Above right *We know that one effect of global warming is a rise in sea level. By planning ahead and taking action now, we can save money and help sustain the global economy in the future. Here are nine possible responses to sea-level rise.*

Far right *Will there be more storms in the future? Will they be stronger? Will they affect different areas than at present? We need to know the answers to questions such as these if we are to plan for the future.*

MODEL RESPONSE STRATEGIES TO DEAL WITH SEA-LEVEL RISE

	BUILDINGS	*WETLANDS*	*CROPS*
RETREAT	*Establish building setback codes*	*Allow wetland migration*	*Relocate agricultural production*
ACCOMMODATE	*Regulate building development*	*Strike balance between preservation and development*	*Switch to aquaculture*
PROTECT	*Protect coastal development*	*Create wetland/mangrove habitat by landfilling and planting*	*Protect agricultural land*

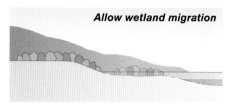

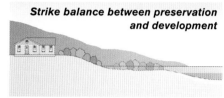
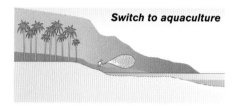

ADAPTING TO CLIMATE CHANGE

The IPCC believes there are six reasons why we must adapt to climate change:

1) Climate change cannot be avoided.

2) Anticipating climate change and taking precautionary measures is more effective and less costly than emergency fixes.

3) Climate change may be more rapid and more pronounced than current estimates suggest, and, as we have seen, unexpected events are more than just possible.

4) We can gain immediate benefits from taking steps to adapt to climate variability and extreme atmospheric events. With hurricane risk, for example, enforcing stricter building laws and establishing better evacuation procedures will leave us better equippped to deal with disasters.

5) We can also gain immediate benefits by abolishing practices that leave us open to increased risk of storm damage, such as building on flood plains and vulnerable coastline areas.

6) Climate change brings opportunities as well as threats. If we approach this in the right frame of mind, future benefits can result from climate change.

The IPCC has provided many ideas on how we can adapt to climate change. The illustration above shows some of the ways in which different countries can adapt to the predicted rise in sea levels.

ADAPTING TO GLOBAL WARMING: THE THREAT TO FRESHWATER

Today, rising human populations, particularly growing concentrations in urban areas, are putting great stress on water resources. The impacts of climate change—including changes in temperature, precipitation, and sea levels—are expected to have varying

PLANNING FOR CHANGE

The major threat from global warming is its unpredictability. Humanity can live in almost any extreme of climate from deserts to the Arctic, but only because we can predict what the extremes of the weather will be. So adaptation is really the key to dealing with the global warming problem–but it must start now, as infrastructure changes can take up to 50 years to implement. For example, if you want to change land use by building better sea defenses or turning farmland back into natural wetlands in a particular area, it can take up to 20 years to research and plan the appropriate changes. It can then take another ten years for the full consultative and legal processes. It can take another ten years to implement these changes and a further decade for the natural restoration to take place.

Money must be invested now. But the problem is that many countries simply do not have the money–and in other parts of the world, people do not want to pay more taxes to fund adaptation. Despite the fact that all of the adaptations discussed will, in the long term, result in financial saving for the local area, the country, and the world, globally we still tend to take a very short-term view.

consequences for the availability of freshwater around the world. For example, changes in river runoff will affect the yields of rivers and reservoirs, and thus the speed at which groundwater supplies recover. An increase in the rate of evaporation will also affect water supplies and contribute to the salinization of irrigated agricultural lands. And, of course, rising sea levels may result in saline intrusion in coastal aquifers, thus contaminating drinking-water supplies.

Currently approximately 1.7 billion people, one-third of the world's population, live in countries that are short of water. The IPCC reports predictions which suggest that, with projected global population increase and climate change and assuming present water consumption patterns, this figure could well increase to five billion people by 2025.

Climate change is likely to have the greatest impact in countries that consume large amounts of water. Regions with already

abundant water supplies will get more water than they want, with a massive increase in flooding. In Europe, for example, computer models predict much heavier rains and thus major flood problems.

Paradoxically, countries that currently have little water (for example, those that rely on desalinization) may be relatively unaffected. It is those countries in between, which have no history or infrastructure for dealing with water shortages, that will be worst hit. In central Asia, North Africa, and southern Africa there will be even less rainfall than there is now, and water quality will become increasingly degraded through higher temperatures and pollutant runoff. Thus terrible droughts in Africa and massive floods in Europe, the United States, and South-East Asia will be the norm in the future. Those countries that have been identified as being most at risk need to start planning now to conserve their water supplies and/or deal with the increased risk of flooding.

KEY

Water withdrawal as % of total available

- *Less than 10 percent*
- *10 –20 percent*
- *20–40 percent*
- *More than 40 percent*

Below *The maps below show which countries had a significant proportion of their population affected by water shortage in 1995, and which countries are likely to be affected by 2025 due to the effects of global warming.*

FRESHWATER STRESS 1995

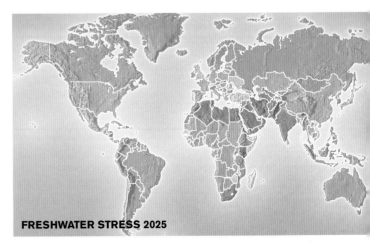

FRESHWATER STRESS 2025

ADAPTING TO GLOBAL WARMING: THE THREAT TO AGRICULTURE

One of the major worries concerning future climate change is the effect it will have on agriculture, both globally and regionally. The main question is, can the world feed itself under the predicted future global warming conditions?

Right A stark vision of the future can be seen in pictures of the dust bowl that occurred in Texas, USA, in the 1930s. In whole areas, the soil was simply blown away.

GROWING COFFEE IN UGANDA

One example of the regional problems that global warming could cause is the case of coffee-growing in Uganda. Here, a temperature increase of 4°F (2°C) would reduce the total land suitable for growing Robusta coffee to less than 10 percent of its current area. Only land at higher elevations would remain suitable; the rest would become too hot to grow coffee. This demonstrates developing countries' vulnerability to the effects of global warming: their economies often rely heavily on one or two agricultural products. Hence one major adaptation to global warming should be to broaden the economic and agricultural base of the most threatened countries. This is, of course, much harder to do in practice than it is on paper.

Below *Many less developed countries rely on cash crops such as coffee to fund everything from industrial development to essential food imports.*

Predictions of cereal production for 2060 suggest that there are still huge uncertainties about whether climate change will cause global agricultural production to increase or decrease. On a worldwide scale, change is expected to be small or moderate. However, this prediction masks the huge changes that will occur at regional level. There will be both winners and losers—with, of course, the poorest countries, which are least able to adapt, being the losers.

Unfortunately, world agricultural production has very little to do with feeding the world's population and much more to do with trade and economics. This is why the EU has stockpiles of food, while many underdeveloped countries export cash crops, such as sugar, coca, coffee, tea, and rubber, but cannot adequately feed their populations.

In the computer models, markets can reinforce the difference between agricultural impacts in developed and developing countries. Depending on the trade model used, agricultural exporters may gain financially even though the supplies fall, because with scarcity the world prices rise.

The other completely unknown factor is how adaptable a country's agriculture can be. For example, the models assume that developing-country production levels will fall more compared with those of developed countries, because their estimated capability to adapt is less than in developed countries. But this is just another assumption that has no analogue in the past.

LEAD TIMES FOR RESPONSE STRATEGIES

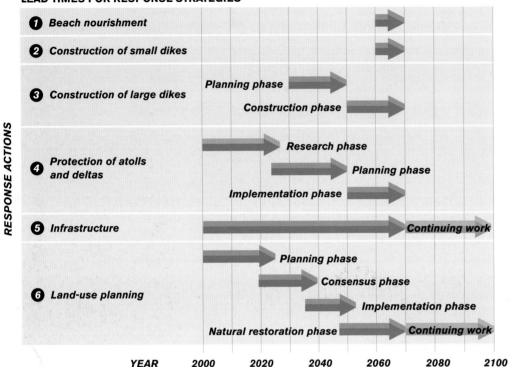

RESPONSE ACTIONS

1 Beach nourishment

2 Construction of small dikes

3 Construction of large dikes
Planning phase
Construction phase

4 Protection of atolls and deltas
Research phase
Planning phase
Implementation phase

5 Infrastructure
Continuing work

6 Land-use planning
Planning phase
Consensus phase
Implementation phase
Natural restoration phase
Continuing work

YEAR 2000 2020 2040 2060 2080 2100

Above *If we are to plan and respond to the threat of global warming, we need to do it now. Many essential changes in the infrastructure and economy of a country can take 50 to 100 years to implement.*

Left *Much of the environmental and social damage of the 1930s' American dust bowl could have been avoided had we known then what we know now. We must be prepared to adapt, and to invest money now in order to save money in the future.*

ADAPTING TO GLOBAL WARMING: THE THREAT OF DISEASE

The transmission of many infectious diseases is affected by climatic factors. Infective agents and their vector organisms are sensitive to factors such as temperature, surface water, humidity, wind, soil moisture, and changes in forest distribution. There is, for example, a strong correlation between increased sea-surface temperature and sea level, and the annual severity of the cholera epidemics in Bangladesh. With predicted future climate change and the massive rise in Bangladesh's relative sea level, cholera epidemics are set to rise.

It is therefore projected that climate change and altered weather patterns will affect the range (both altitude and latitude), intensity, and seasonality of many vector-borne diseases (that is, those carried by another organism, such as malaria, which is carried by mosquitoes) and other infectious diseases. In general, increased warmth and moisture due to global warming will enhance transmission of diseases.

While the potential transmission of many of these diseases increases in response to climate change, we should remember that our capacity to control the diseases will also change. New or improved vaccines can be expected; some vector species can be constrained by use of pesticides. Nevertheless, there are uncertainties and risks here, too: for example, long-term pesticide use breeds resistant strains and kills many predators of pests.

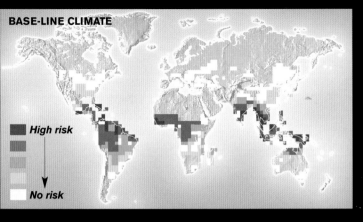

BASE-LINE CLIMATE

High risk

No risk

CHANGE IN EPIDEMIC POTENTIAL

Doubled risk

No change in risk

MALARIA

The most important vector-borne disease is malaria: there are currently 500 million infected people worldwide—about twice the population of the United States. The disease is carried by the Anopheles mosquito.

The main climate factors to have a bearing on the mosquito's malarial transmission potential are temperature and precipitation. The maps above show (left) the current risk of catching malaria and (right) the risk if there were a 0.6°F (1°C) increase in global temperatures in the future.

Assessments of the potential impact of global climate change on the incidence of malaria suggest a widespread increase in risk due to expansion of the areas suitable for malaria transmission. The predicted increase is most pronounced at the borders of endemic malarial areas and at higher altitudes within malarial areas. The local environment conditions, the effects of socio-economic development, and malaria control programs also need to be taken into consideration.

The incidence of infection is most sensitive to climate changes in areas of South-East Asia, South America, and parts of Africa. Global warming will also provide, for the first time ever, the right conditions for mosquitoes to breed in the northern United States, and in Europe.

DENGUE FEVER

Another frightening disease is dengue fever, which affects up to 30 million people each year. A warmer climate increases the transmission of dengue fever. The diagram opposite depicts the number of weeks of potential dengue transmission in the United States at current temperatures (shown in green) and with temperature increases of 4°F/2°C (shown in blue) and 8°F/4°C (shown in red). As you can see, the increases are staggering.

CONCLUSION

It's not all doom and gloom!

The negative aspects of climate change are all too evident. Thankfully, there is also widespread awareness of the problems—and a deep-rooted commitment on the part of many people to put things right.

WHERE THERE'S A WILL, THERE'S A WAY

At the start of this new millennium, we have a greater awareness of the way our planet operates than at any time in the past. This knowledge is invaluable in helping us prepare for the storms and natural disasters that have afflicted humankind since the beginning of time. At the same time we realize that, no matter how difficult it may be, we must take responsibility for the impact that our actions have on the Earth and its climate.

This widespread commitment to change is surely one of the most positive things to come out of the doom-laden forecasts of recent years. Phrases such as "global warming," which were coined to describe very complex, inter-related climatic conditions, are now common currency; and millions of ordinary men and women all over the world are determined to do whatever they can to halt the destruction of our planet before it is too late.

Left *Storms are an essential part of the global climate system: despite the devastation that they cause, they also bring life-producing rains. What we need to do is predict what is going to happen to them in the future, so that we can adapt to cope with them.*

After ten years the scientific case is so strong for global warming that the debate about "will it—won't it happen," seems to be at an end. The 2001 Intergovernmental Panel on Climate Change (IPCC) reports will come to be seen as the turning-point in the whole global warming debate.

One of the most extraordinary sources of support has been big business. With the exception of some United States oil companies, the business community is reacting rapidly to the threat of global warming. In the last five years companies such as Ford and oil companies such as BP and Shell have begun to pour billions of dollars into research into new technologies. Wind power is now mainstream, solar power is in rapid development, hybrid cars are on the roads. Cars that run on fuel cells, hydrogen, and compressed air are no longer pipe dreams.

OUR CHALLENGE

Global warming is one of the greatest threats to face humanity. It is also humanity's greatest challenge. Human ingenuity is such that finding acceptable and affordable alternatives to fossil fuel is well within our current technological grasp. All that is needed is strong global political support. The evidence set out in

Left *Our understanding of hurricanes is increasing, and we are even getting better at predicting how and where they will travel.*

Below *Solar-powered cars are no longer the dreams of science-fiction writers, but could be a realistic alternative to vehicles powered by fossil fuels in the near future.*

SUPPORT AT THE HIGHEST LEVEL

Thanks to the Kyoto Protocol, 186 nations have, for the first time, made a legally binding agreement to cut emissions. The amount agreed may be very small, but at least it is a step in the right direction. Moreover, many experts hope that widespread political pressure will eventually persuade the United States to join the treaty and to take the lead in combating this global threat.

ICELAND—A SHINING EXAMPLE

Let's look at Iceland, which is leading the way to the future. Currently Iceland obtains 99 percent of its electricity from geysers and hydroelectric dams. But it imports 850,000 tons of oil to meet 35 percent of its energy needs, giving Iceland one of the higher per capita carbon emission rates in the world. However, Icelanders are politically committed to becoming the world's first hydrogen economy, cutting greenhouse emissions to zero in the next 30 years.

Their vision is to develop the technology to split water into hydrogen and oxygen and use the hydrogen as a fuel, hence producing no harmful greenhouse gases. When there is a political will and conviction, something can be done about our obsession with a fossil-fuel economy.

Right *Hydroelectric dams have provided us with non-fossil fuel electricity for decades. However, they do have an effect on the local environment, not least because vast areas behind them must be flooded.*

GLOBAL WARMING SOLUTIONS AT LOCAL LEVEL

More good news! While national governments have been taking time to try and thrash out a global deal, local governments and individuals have, for the last ten years, been pushing forward their own solutions.

The driving force behind a lot of these efforts is the Agenda 21 document accepted at the United Nations Conference on Environment and Development at Rio de Janeiro in June 1992. This document stresses the participation of both local agencies and individuals in developing solutions to environmental change, and the importance of strategies to aid development, and sustainability. Most local authorities have policies addressing the key issues in the Agenda 21 document.

An example of this comes from New Hampshire in the United States, where Governor Jeanne Shaheen has facilitated the meeting of local businesses, government, and the environment sector to brainstorm solutions to cut greenhouse gases within the state. The major problem was that, in the past, companies who had voluntarily reduced air pollution had effectively been punished because, with the introduction of the 1990 Federal Clean Air Act Amendments, companies who had already started to clean up their act were given stricter emission reduction targets than companies who were dirtier.

It was decided that New Hampshire would stand by companies that made voluntary reductions, and this was done through a registry of all the reductions made in greenhouse gas emissions. This collective action has resulted in state legislation which was passed in 1999 and has had many benefits; one immediate effect was a significant improvement in local air quality.

These innovative solutions were also noticed by Wisconsin and California and similar legal processes were completed there in 2000. Wisconsin was the first state in the United States to complete a statewide climate action cost study. They found that implementing solutions that cost nothing or even saved money (for example, energy efficiency measures) would create over 8,000 new jobs in the state, saving nearly half a billion dollars. This would also have the effect of raising Wisconsin's state gross product, and would reduce carbon dioxide emissions by over 75 million tons.

Opposite page *Wind-powered generators provide a free and clean means of producing electricity.*

Left *The changing climate affects us all, regardless of which part of the world we live in.*

Below *The technology for solar power for home and industry has reached the point where it is an affordable alternative to fossil fuels. In the future, many of your neighbors' houses—and maybe even your own—will sport solar panels on the roof.*

this book demonstrates that, if we do not react, then the consequences will be disastrous. The impact will be felt around the globe as climate change produces ever more violent and unpredictable weather conditions and the very environment in which we live changes beyond all recognition. As usual, the poorest people in our global society will suffer: their numbers will be measured in billions. But it is not yet too late.

GLOSSARY

AEROSOL Tiny particles of liquid or dust suspended in the atmosphere.

ALBEDO The ability of a surface to reflect light.

ANTARCTIC BOTTOM WATER (AABW) Very cold, salty, deep water which is formed around Antarctic. Because it is so dense, it sinks and flows beneath other deep waters into the Atlantic, Indian and Pacific Oceans.

AQUIFER A store of water trapped in rocks beneath the ground.

ATMOSPHERE The gas layer around the Earth, which extends from the surface to a height of 70 miles (112 km).

ATMOSPHERIC PRESSURE The amount of air that presses down on the Earth. When there is high pressure, a lot of air is pressing down; when there is low pressure, less air is pressing down.

AVALANCHE The rapid movement of large quantities of snow, ice, mud, and rocks down a steep slope.

BLIZZARD A severe winter storm with strong winds, ice, hail, and driving snow.

CIRRUS CLOUD High, white cloud that forms in long, wispy strands.

CLIMATE The general weather conditions in different areas of the world. Climate is sometimes described as the average weather over a period of 30 years.

CLIMATIC, OR WEATHER, FRONT The point at which two air masses with different characteristics meet. When a warm and a cold front meet, then the warm air is forced to rise, thus cooling it down and causing rain.

CORIOLIS FORCE The way in which air and ocean currents are pushed eastward or westward as a result of the spin of the Earth on its axis.

CUMULONIMBUS CLOUD Very large, thick clouds which tower to a great height. They look dark at the base because they are so thick.

EL NIÑO This is a change in the atmospheric pressure over the Pacific which leads to weaker Trade Winds and enables currents of warm water to flow towards South America.

EL NIÑO-SOUTHERN OSCILLATION (ENSO) The climatic system in the Pacific Ocean which varies between the El Niño and La Niña states. These events have a large effect on the weather patterns all around the world.

EQUATOR An imaginary line that runs around the middle of the Earth, halfway between the north and south poles.

EROSION The removal of rocks and soil by the action of wind and water.

FLOOD PLAIN The area, usually flat, alongside a river which is formed by repeated floods.

FUJITA TORNADO INTENSITY SCALE A system of classifying tornadoes on a seven-point scale (F_0–F_6), according to the speed of their winds and the damage they are likely to cause.

GREENHOUSES GASES Gases that are able to trap longwave radiation or heat coming from the Earth, thus warming the atmosphere.

GLOBAL WARMING The effect caused by extra greenhouse gases being put into the atmosphere from industrial pollution, cars, and land-use changes such as deforestation.

GULF STREAM Large warm ocean current which flows from the Gulf of Mexico across the North Atlantic Ocean past Western Europe and into the Nordic Seas.

HADLEY CELLS Three cells of atmospheric circulation which link the Equator to the poles. Within each cell, warm air rises on the lower latitude limb and sinks on the higher latitude limb.

HEMISPHERE Half of the Earth. The Equator divides the Earth into the northern and southern hemispheres.

HUMIDITY The amount of water vapor that is in the air.

HURRICANE A swirling storm in which wind speeds are above 75 mph (120 kph).

INTERTROPICAL CONVERGENCE ZONE (ITCZ) The area where warm moist air masses from the northern and southern hemispheres collide.

JETSTREAM A narrow belt of strong winds blowing at speeds of 100–200 mph (160–320 kph), at a height of 5.5–8.7 miles (9–14 km) above the Earth. Jetstreams form at the Polar Front and the Subtropical Highs.

LATITUDE A way of describing the distance of a particular point from the Equator, measured as an angle from the Equator.

MONSOON The heavy rains that occur at particular times of year in southern Asia and central Africa.

NORTH ATLANTIC DEEP WATER (NADW) Deep water which is formed in the Greenland Sea, Labrador Sea and northern North Atlantic Ocean. This deep water sinks to a maximum depth of 2.5 miles (4 km) and flows southward in the Atlantic Ocean.

PEARSON SCALE A system of classifying tornadoes, according to the length and width of their path.

PERMAFROST Ground that is frozen for large periods of the year due to extreme cold temperatures. In spring and summer in some areas, the top few feet (1–2 m) of permafrost may melt.

PRECIPITATION The movement of water from the atmosphere to the land, in the form of rain, snow, and hail.

STORM SURGE Ocean waves that are enlarged due to the strong winds during a storm. A storm surge can be up to 15 feet (4.5 m) above the normal high-tide level.

TORNADO A violent, spinning column of air.

TRADE WINDS The winds that blow either side of the Equator, from the north-east in the northern hemisphere and the south-east in the southern hemisphere.

TROPICS The hot regions of the Earth that span the Equator, from 30°N to 30°S.

WATER VAPOR Water which is in the air in the form of a gas. It is released as rain or snow when the air cools.

INDEX

Page numbers in *italics* refer to captions.

CREDITS

Quarto would like to thank and acknowledge the following for images reproduced in this book:

Key: t = top, b = bottom, l = left, r = right, c = center,

Allsport: 137br. **ANN RONAN Picture Library:** 134bl. **Cinema Bookshop:** 66b. **GAMMA:** 11cr; 30c; 36tl & bl; 40tr; 50cl; 51tc; 52br; 53tl & bl; 57c & br; 58bl & br; 59br; 62; 63tr & br; 66t; 67tl & br; 69tr; 71tc & br; 75c; 81tr & br; 82bl; 83tl, bl & tr; 84bl & br; 86bl; 87; 99cr; 114t; 118tl; 124; 128br; 129bl; 130tr. **ISI/US Departments of the Interior:** 138br. **NASA (National Aeronautics and Space Administration):** 7r; 8tl (Marit Jentoft-Nilsen, GSFC Visualization Analysis Lab); 11tr (JSC Johnson Space Center); 12tr (produced by Reto Stöckli); 17tl & tc (GSFC); 42l (JSC); 45br (JSC); 57tl; 59t; 64tr (Robert Simmon, Goddard DAAC); 65cr & 65bc (USGS/EROS Data Center); 78tr (GSFC); 98/99c (Landsat 7 Science team/GSFC); 103tr &cr (JSC); 108tl, cl, bcl & bl (NASA/ JPL Jet Propulsion Laboratory/ Caltech); 117bc (Liam Gumley, MODIS Atmosphere Team, University of Wisconsin-Madison/ Cooperative Institute for Meteorological Satellite Studies); 118b/br (NASA/JPL/Caltech); 119tl (NASA/JPL/Caltech); 137tl (Jacques Descloitres, MODIS Land Rapid Response Team NASA/GSFC). **NOAA (National Oceanic and Atmospheric Administration/Department of Commerce):** 1c (Hurricane Research Division); 6l (OAR/ERL/NSSL National Severe Storms Laboratory); 8br (OAR/ERL/NSSL); 9t (OAR/ERL/NSSL); 25tr (Michael Ban Woert, NESDS/ORA); 29tr (Dr. Joseph Golden); 30tr; 30bl, bc & br, 31bl (OAR/ERL/NSSL); 33bl (OAR/ERL/NSSL); 60br, 61tr, bl, bcl, bcr, & br; 67cl (OAR/ERL/NSSL); 72tr (OAR/ERL/NSSL); 74tr, (U.S. Fish & Wildlife Service/J & K Hollingsworth); 75t; 75cr (OAR/ERL/NSSL); 75bl; 75br; 79tl; 79cl (GFDL); 103tc &tcr (GFDL); 107tl; 109tr & cr (Maris Kazmers). **Science Photo Library:** 36tc (Claude Nuridsany/Marie Perennou); 77cr (Claude Nuridsany/Marie Perennou); 127tr; 128tl (Simon Fraser/Mauna Loa Observatory). **SPECTRUM Colour Library:** 55cr. **TRIP Photo Library:** 29br; 32; 37; 39bl; 41br; 44tr; 44br; 45tr; 51tl; 53tr; 55tl; 69b; 70br; 74br; 77tl & tr; 82tl; 86tl & br; 109bl; 111cr & br; 122bl.

All other illustrations are the copyright of Quarto. While every effort has been made to credit contributors, we apologize should there have been any omissions or errors.

The author would like to thank the following: Johanna Maslin; Virginia Ettwein; UCL Department of Geography Drawing Office (D'alton, McBayand Quinn); Environmental Change Research Centre, Department of Geography, UCL.

USEFUL WEBSITES

• National Oceanic and Atmospheric Administration: http://www.noaa.gov/
• Climate Ark–The Premier Climate Change & Renewable Energy Portal: http://www.climateark.org/
• UK Met Office (weather and climate): http://www.metoffice.com/weather/index.html
• UCL Environmental Change Research Centre: http://www.geog.ucl.ac.uk/ecrc/

FURTHER READING

• Harvey D., *Global Warming: The Hard Science*, Prentice Hall, 2000
• Houghton J.T. et al (editors), *IPCC, 2001: Climate Change 2001: The Scientific Basis*, Cambridge University Press, 2001
• Maslin M., *Global Warming: Causes, Effects and the Future*, Voyageur Press, 2002
• National Assessment Synthesis Team *Climate Change Impacts on the United States, Overview*, Cambridge University Press, 2000